Extinction and Phylogeny

Extinction and Phylogeny

Michael J. Novacek and
Quentin D. Wheeler

EDITORS

Columbia University Press
New York

Columbia University Press
New York   Oxford
Copyright © 1992 Columbia University Press
All rights reserved

Library of Congress Cataloging-in-Publication Data

Extinction and phylogeny / editors, Michael J. Novacek and Quentin D.
    Wheeler.
        p.   cm.
    Includes bibliographical references and index.
    ISBN 0-231-07438-7
    1. Extinction (Biology).   2. Phylogeny.   I. Novacek, Michael J.
II. Wheeler, Quentin.
QE721.2.E97E967   1992
575.7—dc20   92—5856
                                                     CIP

Casebound editions of Columbia University Press books
are Smyth-sewn and printed on permanent and durable acid-
free paper.

Printed in the United States of America

c 10 9 8 7 6 5 4 3 2 1

Contents

Extinction and Phylogeny

Introduction

Extinct Taxa: Accounting for 99.999 . . . % of the
Earth's Biota

Michael J. Novacek and Quentin D. Wheeler

Systematics faces a daunting agenda. Not only must it provide a vocabulary
for the organization of the living flora and fauna, but it must account for the
sources of histories of the component species. The influence of history here is
profound—patterns of relationships for myriad extinct forms are obviously
interwoven with those patterns among living taxa.

Consider the sheer weight of the historical component. Life on this planet
has existed for at least 3.5 billion years, a history marked by a series of major
diversification events. These include the appearance of eukaryotes (about 1.9
billion years ago); the proliferation of multicellular lineages, especially in the
marine realm (about 700 million years ago); and colonization of the land by
bryophytic plants, terrestrial arthropods, and many other groups (about 450
to 375 million years ago). Although the present global biological diversity is
probably at an all-time high (Wilson 1988), marked species' abundance must
pertain to a significant span of geological time. Plant diversity, for example,
has been increasing appreciably since about 700 million years ago, and
increasing very steeply since about 100 million years ago (Knoll 1986; Crepet
et al. 1991). Such a proliferation is evident despite a series of mass extinction
events (Raup and Sepkoski 1984). A comparison of these trends against
estimates for the average longevity of species is illuminating. Although such

estimates are tentative, particularly for terrestrial organisms, a range of longevity between 1 and 10 million years for most species seems broadly acceptable (Raup 1984). Despite the relatively shorter spans of individual species, high levels of diversity have persisted over hundreds of millions of years. This marked turnover in species requires an enormous portion of extinct taxa. Indeed, there is general agreement that more than 99 percent of all the species that ever existed have become extinct (Raup 1979, 1981).

These patterns of diversification bring into focus a critical question: How profoundly does the history of past life color our understanding of the evolution and relationships of living species and higher taxa? Unfortunately, the evidence of extinct life—the fossil record—cannot provide all the empirical data bearing on this question. The record displays instead only a sprinkling from the vast pool of former diversity. This deficiency is obvious and almost tritely emphasized. In fact, the curve for recognition of the importance of fossil evidence has gone through several cycles. For much of this century, it was popular to regard phylogenetic analysis as seriously biased, even unjustified, without a reasonably good fossil record. In recent decades, however, fossil information has been impugned with increased frequency on a number of fronts. Some cladists have said that all data should be analyzed in the same ways and that fossils are inferior to neontological data because of the "selective" way in which fossilization preserves only a limited number of characters (Hennig 1966, 1981; Patterson 1981). Molecular biologists have spoken of a level of precision in their data that they claim is unmatched in comparative morphologic data in living species, let alone in imperfectly preserved fossils (Goodman 1989). And the long-standing debate over the adequacy of the fossil record in recording real evolutionary patterns was refueled by controversies over the origin of species (Eldredge 1971; Eldredge and Gould 1972).

In spite of these controversies, no one has denied that fossils provide a window on the past or that unique pieces of information about character combinations, ancient taxa, minimum absolute ages, and former geographic distributions can be retrieved only from the fossil record. Accordingly, the "paleontological method" has emerged as one of three commonly cited approaches to formulating hypotheses of character polarity that, along with outgroup comparisons (Watrous and Wheeler 1981; Maddison et al. 1984) and ontogeny (Nelson 1978; Humphries 1988; Wheeler 1990a), form a modern rendition of Agassiz's threefold parallelism (Nelson and Platnick 1981).

Some of the critics, in retrospect, have underestimated the potential role for fossils in their analyses. It is clearly undesirable to exclude invaluable fossil evidence. It is equally undesirable, however, to permit an incomplete

fossil data set to detract from resolution available from comparative neontological studies.

The fossil record thus provides a clear, albeit incomplete, access to extinct
lineages and their bearing on the phylogeny of all life. Extinction can be
viewed even more broadly as the loss of information relevant to studies of
living groups as well as fossils. Just as paleontological research uncovers new
and important taxa, discoveries of new taxa are part of our exploration of the
present-day tropics, the ocean basins, and other frontiers of the biological
world. Our inability to document these taxa adequately and expediently is a
critical problem (Wilson 1988). Many of these species are going to be extinct
before we have a chance to account for them. Global deforestation is imposing
a human-mediated rate of extinction believed to be accelerated several hundred
times over the natural rate in recent geological time (Wilson 1985, 1988). It
is important, then, to determine how any shortcoming in taxonomic sampling
affects our ability to characterize biodiversity and phylogenetic history accurately.

The biological diversity crisis is, in fact, converting these problems from
the abstract to the concrete (e.g., Raup 1988). Expansive tracks of forest are
being destroyed or altered irreversibly every year, and authorities agree that
30–50% of all living species may face extinction in the next three to five
decades (Guppy 1984; Myers 1984; NSB 1989; Wilson 1985, 1988; Wheeler
1990b). Whatever the ill effects of extinction are on our ability to reconstruct
evolutionary history, these effects will be exacerbated by this runaway rate of
extinction. In the next few decades, scientists will have to selectively sample
countless species of flowering plants, insects, vertebrates, microbes, fungi,
and other organisms, especially in the "lower" plant and invertebrate animal
groups, and selectively preserve them in natural or artificial habitats. Fossil
deposits must also receive increased attention because of accelerated rates of
human disturbance of sites. Such decisions will likely determine the range of
phylogenetic information available to future generations of systematic biologists. Without knowledge of phylogenetic relationships and without a firm
grasp of the impacts of extinction, how are such decisions to be intelligently
made?

Scope and Organization

This book assembles original work concerned with the role of extinction in
phylogenetic analysis. Some chapters naturally deal with the primary database
for extinct organisms—the fossil record. Discussion also leads to a consideration of conserved specimens of species facing imminent extinction. Other

chapters are concerned with the recovery of information from the living biota, taking into account the limitations of sampling. Complementary discussions involve the analysis of data sets of variable completeness (for example, partially preserved fossils or patchy samples of extant taxa).

In dealing with these questions, we can claim progress beyond some previous debates. For example, it would seem a sterile exercise merely to decide whether fossils are important to phylogenetic study. The obvious answer is that fossils, like extant taxa, provide a wealth of relevant information and should not be ignored. At the same time, the study of fossils is not in all respects comparable to the study of living taxa; the temporal dimension provided by the fossil record requires distinctive methods of analysis. Moreover, the comparison of fossils to extant forms may pose problems concerning missing data in matrices. A general review of the parameters for age-dating fossil occurrences and the phylogenetic patterns revealed by the extraordinary fossil record of horses is provided by B. J. MacFadden (chapter 1).

We do know that evidence of extinct forms has significantly altered phylogenetic hypotheses of some living taxa, whereas in other instances such evidence has had little impact (Donoghue et al. 1989). The related question is, "Under what conditions would we expect to see such a pivotal role for fossil data?" An analytical approach here is to resolve a tree of relationships based strictly on data from living taxa ("the Recent tree"). One can then add fossil taxa and measure the degree to which the relationships among the original extant groups are reoriented (e.g., Gauthier et al. 1988). An elaboration of this approach and some related issues are discussed by M. J. Novacek (chapter 2).

Another issue concerns the use of phylogenies as a framework for historical questions. Relevant here are the potential insights gained from fossil distributions concerning the study of evolutionary rates. M. A. Norell (chapter 3) discusses some of the potential shortcomings in the use of fossil age data to retrieve patterns consistent with cladograms based purely on characters (and not on time of occurrence). Norell proposes some original ways for estimating rates of taxonomic change and diversity that account for the error in fossil dates for the origin of taxa.

One of the central questions in studies of past and present diversity concerns the concept of a species and its accessibility to phylogenetic analysis. Are species reproductively isolated units, lineages with unique histories, definable monophyletic groups, diagnosable clusters that defy subdivision, or something else? Authors K. C. Nixon and Q. D. Wheeler (chapter 4) expand on the phylogenetic species concept, focusing on the importance of extinction events in the origin of species.

Two succeeding chapters emphasize the empirical aspects of case studies in order to illustrate general principles. G. D. Edgecombe (chapter 5) draws on evidence from Early Paleozoic trilobites to demonstrate how a modern phylogenetic study can force serious revision of popular notions concerning mega-evolutionary patterns. D. A. Grimaldi (chapter 6) applies his studies of tsetse flies to the argument that fossils and past distributions have a critical and very clear operational bearing on the reconstruction of biogeographic histories.

Some of the issues and methods discussed in the volume have a strong relevance to more recent investigations of extant groups, such as molecular systematics. W. C. Wheeler (chapter 7) relates simulation models to the issue of limitations and incompleteness—namely, when is information sufficient to uncover meaningful patterns even though the available data are variably complete with respect to taxa and characters? And what are the effects of increased taxonomic, as opposed to increased character, sampling? This inquiry not only impacts on the morphologic study of fossils and extant taxa, but also has clear implications for molecular work, where direct evidence of "extinct DNA sequences" and the like is unavailable for the overwhelming majority of fossil taxa.

A final treatment in the book, contributed by K. C. Nixon and Q. D. Wheeler (chapter 8), ties many of the foregoing issues to conservation problems and the biodiversity crisis. Since priorities must be set for conserving the living world (we cannot or will not save all species), it is important to have a sense of *true* diversity—namely, diversity that reflects phylogenetic history rather than simply a convenient subdivision of kinds. It is also important to identify species that have a particular role, by virtue of their phylogenetic position, in revealing a closer approximation to this true diversity. As we might expect, such a procedure depends on credible phylogenies that could draw on both fossil and extant evidence.

Although this sequence of chapters follows a logical pathway, it is not rigorously fashioned. The contributions are meant to be freestanding; a theme in one chapter does not directly depend on thoughts developed in another. Nonetheless, readers will find that the expected overlap in some topics—such as the relationship between fossil age and primitiveness—is worthy of discussion from different perspectives. Readers will also note contradictory views on such matters as anagenesis (transformation from one taxon to another without splitting) and the definition of species. We hope that juxtaposition of these views stimulates further interest in some clearly challenging questions.

Some Essentials of Cladistic Analysis

Although their approaches vary, all contributors to this book share an understanding and interest in cladistic methodology. This conformity links the different issues and examples found in the chapters. Many discussions focus on newly emerging, and somewhat technical, issues in cladistics. The book is intended, however, not just for systematists (including paleontologists) but for a broader audience of informed readers interested in problems involving fossils, past and extant diversity, extinction, and phylogeny. This brief description of basic cladistic methods provides some background information.

Modern cladistics is rooted in Willi Hennig's (1966) "phylogenetic systematics" in that it stresses the recognition of *synapomorphy* (shared-derived characters) as the basis for close (sister-group) relationship. The emergence of cladistics in the 1970s was controversial, but this method is now popular among paleontologists and neobiologists (including molecular biologists). Certain kinds of data (e.g., morphometrics, immunological distances, or DNA–DNA hybridization) are distance measures not easily transformed into discrete characters. These data are more amenable to *phenetic* clustering based on overall similarity. Such techniques often have the disadvantage of assuming uncertain evolutionary processes (for example, the constant or clocklike rates of nucleotide change in some DNA hybridization analyses) to achieve branching patterns that make biological sense.

It is important to emphasize that many molecular studies now deal with discrete character data in the form of amino acid replacements in proteins or base substitutions in DNA sequences (primarily from ribosomal or mitochondrial genes). Such studies use the same cladistic approaches that pertain to morphologic studies. Nonetheless, it is often claimed that molecular evidence is superior for phylogenetic reconstruction because it is a more direct reading of genetic and genealogical change, without the trappings of biased interpretation of morphologic structure and function. Moreover, molecular data, consisting simply of hundreds of nucleotides coded in a data matrix, are not subject to the "sanitization" procedure necessary to distill and organize other kinds of evidence. Molecular studies, however, are sometimes plagued with equal or greater inconsistencies in results than those seen in other kinds of studies (Wyss et al. 1987; Novacek et al. 1988; Faith and Cranston 1991). Some recent analyses of nucleotide sequences, such as that for ribosomal DNA, reveal that primary structure (nucleotide sequences) is in itself not necessarily adequate for phylogenetic reconstruction and that secondary or tertiary structure of the molecule must also be taken into account for such purposes (Wheeler and Honeycutt 1988). There is thus a move toward more interpretive study of molecular structure and function that converges on approaches taken in morphologic work. It is further noted that morphologic data may be reexamined in finer detail (microstructure, ontogeny, and ultimately molecular), while nucleic acid bases represent an elemental level of structural identity not subject to additional "testing."

The issue of character interpretation is of central interest. Although cladistics is useful as a logical framework for constructing patterns of relationship from distributions of synapomorphies, it does not establish the methods by which we choose and describe characters in the first place. There is no clear rule for how many attributes are recognized or correlated with each other or for what kind of attribute is considered. The development of a character table for the purpose of cladistic analysis remains a difficult and not altogether objective exercise. The only means of judging the effectiveness of this stage of research, often referred to as "the character analysis," is to examine the stability of cladistic patterns derived from the characters. For example, a worker may decide to construct a cladogram for higher mammals based on hair color, and there is nothing beforehand that prohibits this exercise. Yet such a set of characters might be very poor for indicating mammalian relationships—little resolution will be achieved and many contradictory patterns will come from other kinds of character evidence. By contrast, character information from the mammalian skull and reproductive system has proven relatively useful in identifying certain higher groups of mammals. An accounting of the widest range of characters possible and continued expansion of the data set to accommodate new character evidence are the recommended procedures.

In cladistics the only groups considered natural are *monophyletic,* that is, groups that include an ancestor and all its descendants. Synapomorphies provide the only evidence for recognizing such groups. This is because a given set of synapomorphies represents a unique combination of attributes that does not diagnose more than one group. The *combination* of hair, mammary glands, and three ear ossicles does not uniquely diagnose any group of vertebrates other than mammals, nor does the *combination* of a modified trophoblast, definitive chorioallantoic placenta, prolonged intrauterine gestation, corpus callosum in the brain, and reduced molar stylar shelves diagnose any group of mammals (or, for that matter, vertebrates or organisms) other than Eutheria (placental mammals).

Opposed to this strategy would be the recognition of groups by their lack of such unique sets of attributes. For example, we might recognize a marsupial plus monotreme (duckbilled platypus and echidna) group by the absence of the characters cited previously for eutherians. Such a grouping, however, would be inconsistent with cladistic logic unless marsupials and monotremes could be united by a set of synapomorphies that excluded eutherians, and if there were no stronger contradictory evidence that linked eutherians with either marsupials or monotremes. Such contradictory evidence does exist, of course. This evidence groups eutherians and marsupials within Theria. Under the presently accepted scheme, marsupials plus monotremes do not form a monophyletic group but a *paraphyletic* group. They constitute a group with an ancestor but only some of its descendants.

Paraphyletic groups are common in the systematic literature (for example, the traditional "reptiles," "invertebrates," "archeobacteria," "proteutheria"), and some persist for reasons of convenience in the absence of systematic knowledge. In cladistic studies, paraphyletic groups are not recognized even for purposes of convenience or stability. Some workers object that classifications that only incorporate monophyletic groups are too explicit, and thus are likely to be continually altered. They also object to the great increase in names and categories that such "phylogenetic classifications" require (Mayr 1974). These criticisms are practical rather than substantive in nature. A classification is most effective if it reflects meaningful phylogenetic information. Here such meaningful phylogenetic information is equated with the hierarchy of monophyletic groups. And phylogenetic classifications, like other hypotheses, are not inherently unstable but change in response to subsequent tests.

Since monophyletic groups are diagnosed by sets of synapomorphies, or shared–derived attributes, the determination as to whether a character is derived (*apomorphic*) or primitive (*plesiomorphic*) is critical to cladistic analysis. For this operation some general procedures have been identified. *Outgroup comparison,* the most widely used procedure, relies on the logical argument that a character state found in the relatives of a group (the outgroup) is likely also to be the primitive state for the group in question—the ingroup (Watrous and Wheeler 1981, Maddison et al. 1984). This rule applies even when the states for the character may vary within the ingroup. As a simple example, we note that some mammals (monotremes) lay eggs from which the young are hatched (ovoviviparity), whereas in other mammals the eggshell is never deposited, and the young emerge directly from the genital opening (viviparity). Thus within mammals (the ingroup), there are two states for this reproductive condition. Outgroup comparison suggests the egg-laying condition is primitive (plesiomorphic) for mammals. Although we do not know the reproductive mode of the extinct "therapsids," which are the nearest relatives to mammals (Gauthier et al. 1988), we do know that living reptiles, birds, and other tetrapods are almost entirely egg laying (some snakes and lizards are notable exceptions in being "live-bearers"). The weight of the evidence favors the conclusion that some mammals simply retain the primitive egg-laying (ovoviparous) state and that the development of viviparity was a secondary event. Viviparity is, in fact, part of the evidence for the monophyly of Theria.

In many cases, the primitive state for a taxon can be ambiguous. The primitive state for the ingroup can only be determined if the primitive states for the nearest outgroup are easy to identify and these states are the same for at least the two nearest outgroups (Maddison et al. 1984). If live bearing and egg laying varied markedly across many groups of tetrapods, including the nearest relatives of mammals, then we might not be able to identify the primitive state for mammals.

From the preceding it is clear that outgroup comparison relies on a previously established pattern of relationships that includes both the ingroup and its nearest relatives. The development of this framework also depends on the identification of states based on outgroup comparison, which refers to a yet more general level of branching patterns. Thus we can see the analysis of characters extending ever outward to the corners of biological diversity. An outgroup analysis is only as reliable as the higher-level pattern upon which it is based. Accordingly, outgroup comparison has been called an "indirect" means of character evaluation (Nelson 1978). This may be viewed as a shortcoming if there are completely independent lines of evidence for whether a trait is primitive or derived.

"Direct" approaches are those supposedly not requiring a previously established pattern of relationships. Two "direct" approaches have been widely advocated. These are the *ontogenetic* and *paleontological approaches*. Ontogeny is a powerful means of identifying the status of characters. There seems to be an overwhelming number of examples where early stages of ontogeny can be equated with generally distributed states (equals phylogenetically primitive) and later ontogenetic stages can be equated with highly specialized states (equals phylogenetically derived) (von Baerian recapitulation—see Patterson 1983; Kraus 1988; Mabee 1989; Humphries 1988; Wheeler 1990a). This relationship has exceptions. Various ontogenies may actually show late stages that equate more with early embryonic stages in their relatives (paedomorphosis). The ontogenetic pathway in one organism is not in itself sufficient to determine whether the character in question is primitive or derived (Kluge 1985). And what is directly measured appears to be "character adjacency," not polarity (Wheeler 1990a; Nixon and Wheeler, in preparation). More effective is a broader-based comparison among the ontogenies of organisms (Fink 1982), wherein ontogenies themselves are treated as one would treat characters in outgroup comparison (de Queiroz 1985). In such comparisons, the labeling of the ontogenetic approach as a "direct method" of character analysis is thus misleading (Wheeler 1990a).

As a direct means of assessing characters, the paleontological approach has proven even more controversial than the use of ontogeny. The paleontological approach follows the general rule that older fossils show more primitive characters than younger fossils and Recent organisms. Coincident with the rise of cladistics is a strong criticism of this logic, a criticism even expressed by paleontologists (Schaeffer et al. 1972). Gaps in the fossil record can plague the use of the paleontological rule, and there is clearly no law that equates age with primitiveness. Moreover, differential extinction patterns may result in more recent lineages becoming extinct at earlier times (Eldredge and Cracraft 1980). There is, however, strong expectation that more ancient fossils, as a general rule, preserve relatively primitive states (Simpson 1975; Eldredge and Novacek 1985). This is shown in studies (Gauthier et al. 1988) that take into account the cladistic patterns that incorporate both fossil and living taxa. Clearly, one of the outstanding features of fossils is that they may preserve character states marked by extreme transformation in living taxa. Accordingly, Gauthier et al. (1988) found that 15% of the characters bearing on the relationships of mammals with other tetrapods were essentially missing because of the extreme transformation in Recent mammals. Modifications of the quadrate, for example, which link some mammals with some, but not all, synapsids, cannot be determined in living mammals. This is because in all living mammals the quadrate has transformed into an incus, a component of the mammalian ossicle system. For this reason, fossils, where present, can play a crucial role in determining patterns of relationships of extant taxa. The point is discussed at some length in chapter 2 of this volume.

Modification of a character refers to a phenomenon known as *character transformation*. We do not customarily observe character transformation in real time (the exception may be refined ontogenetic experiments), but such transformation often can be readily inferred. For systematists, character transformation can be coded in various ways. In some studies all characters are coded in *binary* fashion. For example, the egg-laying condition can simply be scored as primitive (O) for mammals, and the live-bearing condition can be scored as derived (1). There is some reason to argue that two or more derived states may actually be different expressions of a single character or character system. This might lead to a proposal for a sequence of states, one derived from another, in an *ordered multistate transformation*. Such transformations have been used in the case of variation in digit numbers or scale arrangements. Nonetheless, it is often unclear whether any one of a number of states in a proposed transformation can be derived from any other. Because of this ambiguity, multiple states of a given character are often left *unordered* for purposes of cladistic analysis. It may seem that many characters are justifiably coded as multistate characters (e.g., gradations in hair color or vibrissae number), but it is noteworthy that a larger number of multistate characters can make the cladistic analysis a very complicated and time-consuming procedure even with the aid of computer programs (Swofford 1985). Moreover, multiple states are often nothing more than unresolved binary characters.

So far we have considered only the recognition of characters and their various states as either primitive or derived. Once a pattern of distribution for derived characters is accepted, there remains the problem of cladogram construction. The preferred method at this point is *parsimony analysis*. Parsimony yields the simplest or most efficient summary of the character data. It thus yields a scheme of branching relationships that requires the fewest steps (character transformations) to explain the characters (Farris 1983). The distribution of certain characters may not be strictly coincident with the most parsimonious cladogram. The cladogram, however, should show enough internal consistency to withstand some contradiction in the form of characters that suggest alternative groupings. Characters (or character states) whose distributions conflict with the branching pattern of the most parsimonious tree are referred to as *homoplasious characters*. A measure of the information content of the cladogram is the *consistency index* (CI) (Kluge and Farris 1969), which equals the number of character states divided by the total number of steps needed to yield the cladogram. Intuitively, cladograms with high values for CI (greater than 0.80) provide very efficient summaries of the character evidence. It is noteworthy that high consistency is generally harder to achieve with problems involving more biological diversity, and there is an inverse relationship between the consistency index and the number of terminal taxa considered (Sanderson and Donoghue 1989). One limitation of the consistency index is that it may mask instances where many of the state changes on the cladogram are homoplasies that contribute no essential information on branching. This is because each character state change reduces the proportion of similarities that can be recognized as synapomorphies. For this reason, a *retention index* (RI) has been recommended (Farris 1989) for disclosing the fraction of the apparent synapomorphy in the character(s) that is retained as a synapomorphy on the tree. Farris (1989) has provided a description of these and other indices and their application.

The parsimony approach thus offers a way of building schemes of relationship from character evidence. If such schemes are interpreted as the preferred phylogeny, or genealogical history, then any contradictory character evidence, or homoplasy, is judged to be either an instance of convergence or reversal. These secondary events are obviously misleading indicators of close relationship. Cladograms that allow a high degree of homoplasy have low consistency and retention indices. One might predict that such schemes are very unstable and are easily overturned as new data are incorporated. In some instances, however, cladograms with relatively large homoplasy content account for complex data sets and their basic topologies are not strongly altered when additional traits are considered (Novacek 1986).

Since cladistic problems vary in their complexity, the parsimony approach is not always equally effective. Ideally, one would prefer constructing the cladogram from a straightforward manual arrangement of the characters. Often complex data sets require the assistance of computer programs for computations. For example, a tree with only 11 taxa may yield 654,729,075 possible bifurcating (rooted) arrangements (Felsenstein 1978a). If such a tree is to be evaluated with reference to many characters, one can see the need for computers. Two microcomputer programs, PAUP (D. A. Swofford 1985) and HENNIG 86 (J. S. Farris 1988), are particularly effective for such computations. Both these programs have branch-and-bound algorithms that guarantee the discovery of the most parsimonious tree. The use of "branch-and-bound" (or, in HENNIG 86, "implicit enumeration") is usually limited to less than 25 terminal taxa, although the actual limit is a function of computer power and speed. More complex data require enormous computational effort. For such cases, a "heuristic" algorithm is applied, wherein branch swapping proceeds until some stable point in the computation of the shortest tree is reached. Although heuristic algorithms do not guarantee the discovery of the absolute minimum-length tree, they often yield trees identical to those produced by exhaustive tree searches (such as branch-and-bound). HENNIG 86 and PAUP have very effective heuristic search programs. Other computer programs—for example McCLADE (Maddison and Maddison 1987) and ClaDos (Nixon 1988, 1991)—provide a means for manual character tracing and excellent graphic displays of cladograms. McCLADE can be used in conjunction with the Apple MacIntosh version of PAUP (version 3.0) as a tool for inspecting the parameters of most parsimonious trees, whereas ClaDos may be used in conjunction with HENNIG 86 on IBM compatible computers.

The widespread use of the preceding programs indicates that parsimony is now commonly regarded as a very effective logical framework for systematic analysis. Nevertheless, objections have been raised to the use of parsimony in this fashion. It is conceivable that certain historical processes may not yield a genealogical scheme consistent with the most parsimonious branching pattern. For this reason, some workers (Felsenstein 1978b) advocate the use of *maximum likelihood* estimates that derive trees from a given set of conditions meant to represent certain evolutionary processes. The tree that best fits the pattern predicted by these assumptions is the maximum-likelihood tree. On the other hand, we have no strong empirical basis for determining a given set of conditions; we can only surmise what such a set of conditions might be. Thus there is no way to discredit the general application of parsimony; the latter is simply a means of finding the most efficient summary of a given distribution of character evidence.

There are, however, cases where the straightforward application of parsimony can be misleading in the face of empirical character evidence. For example, we may have an independent means of knowing that some character changes occur more readily and more frequently than others. In such cases, it would be unjustified to undertake a search based on parsimony without a weighting factor that accounts for the variances in the probability of character change. Characters that change readily would be downweighted because they are more susceptible to independent origin or reversal and would be less likely to represent monophyletic groups. Such a weighting procedure would likely reduce the amount of homoplasy in the shortest calculated tree. These remarks are issued with the qualification that each character transformation is a unique (evolutionary) historical event and, as such, is not subject to any kind of probability statement.

In reality, we have few examples where knowledge is sufficient for such a priori weighting. A notable case is the observation that transitions (exchanges in DNA sequences that substitute one purine for another or one pyrimidine for another) occur much more readily than transversions (exchanges that substitute a pyrimidine for a purine or vice versa). Algorithms have been developed to accommodate this variance in the rate of nucleotide substitution (Lake 1987). Other programs do provide for character weighting "after the fact" or a posteriori. In HENNIG 86 there is a procedure that increases the weight of characters that show higher-consistency indices *after* the first pass through the data and an unweighted parsimony analysis is completed. *Successive weighting* of these characters may eventually yield a more resolved branching scheme than the unweighted tree (Carpenter 1988).

The successive weighting procedure must be applied with some prudence, as characters may be artificially highly consistent, and thus "up weighted" simply because they are not known for all the terminal taxa. This problem stems from the fact that the algorithms used in parsimony programs account for missing data in a particular way. They assign a character state to the missing entry that conforms to the most parsimonious solution based on known characters (Nixon and Davis, 1991). Accordingly, the effects of missing data are of particular concern in three cases: (1) when fossil taxa are considered (Gauthier et al. 1988; Rowe 1988; Novacek 1989), (2) when extant taxa are so highly transformed that certain character information is essentially missing (Gauthier et al. 1988), or (3) when there are notable deficiencies in sampling of nucleotide substitutions for genes (or amino acid replacements for proteins) in different taxa (Miyamoto and Goodman, 1986; Novacek et al. 1988).

Even when we account for missing data, taxa with incomplete character information can introduce much homoplasy and uncertainty into the analysis. For example, the addition of several extinct eutherian clades represented by incompletely preserved fossils greatly increases the number of equally most parsimonious solutions and reduces the overall resolution of the branching scheme (see chapter 2).

The preceding comments only highlight some of the issues in cladistic analysis. For additional information readers are referred to texts by Eldredge and Cracraft (1980), Wiley (1981), Nelson and Platnick (1981), Schoch (1986), as well as recent issues of the journals *Systematic Zoology* and *Cladistics*. The field of systematics is now enjoying some resurgence in the United States and cladistic methods are today a basic element of college courses taught in systematics.

REFERENCES

Carpenter, J. M. 1988. Choosing among multiple equally parsimonious cladograms. *Cladistics* 4:291–296.

Crepet, W. L., E. M. Friis, and K. C. Nixon. 1991. Fossil evidence for the evolution of biotic pollination. *Trans. Roy. Soc. Lond.* In press.

Donoghue, M. J., J. A. Doyle, J. Gauthier, A. G. Kluge, and T. Rowe. 1989. The importance of fossils in phylogeny reconstruction. *Ann. Rev. Ecol. Sys.* 20:431–460.

Eldredge, N. 1971. The allopatric model and phylogeny in Paleozoic invertebrates. *Evolution* 25:156–167.

Eldredge, N., and J. Cracraft. 1980. *Phylogenetic Patterns and the Evolutionary Process.* New York: Columbia University Press.

Eldredge, N. and S. J. Gould. 1972. Punctuated equilibrium: an alternative to phyletic gradualism. In J. W. Schopf, ed., *Models in Paleobiology,* pp. 82–115. San Francisco: Freeman, Cooper.

Eldredge, N. and M. J. Novacek. 1985. Systematics and paleobiology. *Paleobiology* 11:65–74.

Faith, D. P. and P. S. Cranston. 1991. Could a cladogram this short have arisen from chance alone? On permutation tests for cladistic structure. *Cladistics* 7:1–28.

Farris, J. S. 1983. The logical basis of phylogenetic analysis. In N. I. Platnick and V. A. Funk, *eds.,* Advances in Cladistics 2:7–36. New York: Columbia University Press.

Farris, J. S. 1988. *HENNIG 86.* Version 1.5. Distributed by the author. 41 Admiral St., Port Jefferson Station, N.Y.

Farris, J. S. 1989. The retention index and the rescaled consistency index. *Cladistics* 5:417–419.

Felsenstein, J. 1978a. The number of evolutionary trees. *Syst. Zool.* 27:27–33.

Felsenstein, J. 1978b. Cases in which parsimony or compatibility methods will be positively misleading. *Syst. Zool.* 27:401–410.

Fink, W. 1982. The conceptual relationships between ontogeny and phylogeny. *Paleobiology* 8:254–264.

Gauthier, J., A. G. Kluge, and T. Rowe. 1988. Amniote phylogeny and the importance of fossils. *Cladistics* 4:105–209.

Goodman, M. 1989. Emerging alliance of phylogenetic systematics and molecular biology: a new age of exploration. In B. Fernholm, K. Bremer, H. Jornvall, eds., *The Hierarchy of Life,* pp. 43–61. Amsterdam: Elsevier.

Guppy, N. 1984. Tropical deforestation: a global view. *Foreign Affairs* 62:928–965.

Hennig, W. 1966. *Phylogenetic Systematics*. Urbana: University of Illinois Press.

Hennig, W. 1981. *Insect Phylogeny*. New York: John Wiley.

Humphries, C. J., ed. 1988. *Ontogeny and Systematics*. New York: Columbia University Press.

Kluge, A. G. 1985. Ontogeny and phylogenetic systematics. *Cladistics* 1:13–27.

Kluge, A. G. and J. S. Farris. 1969. Quantitative phyletics and the evolution of anurans. *Syst. Zool.* 18:1032.

Knoll, A. H. 1986. Patterns of change in plant communities through geologic time. In J. Diamond and T. J. Case, eds., *Community Ecology*. Philadelphia: Harper & Row.

Kraus, F. 1988. An empirical evaluation of the use of the ontogenetic polarization criterion in phylogenetic inference. *Syst. Zool.* 37:106–141.

Lake, J. A. 1987. A rate-independent technique for analysis of nucleotide sequences: evolutionary parsimony. *Mol. Biol. Evol.* 4:167–191.

Mabee, P. 1989. An empirical rejection of the ontogenetic polarity criterion. *Cladistics* 5:409–416.

Maddison, W. P., M. J. Donoghue, and D. R. Maddison. 1984. Outgroup analysis and parsimony. *Syst. Zool.* 33:83–103.

Maddison, W. P. and D. R. Maddison. 1987. *MacClade, Version 2.1. A phylogenetic computer program*. (Distributed by the authors.)

Mayr, E. 1974. Cladistic analysis or cladistic classification? *Zeit. Zool. Syst. Evol. Forsch.* 12:94–128.

Miyamoto, M. and M. Goodman. 1986. Biomolecular systematics of eutherian mammals: phylogenetic patterns and classification. *Syst. Zool.* 35:230–240.

Myers, N. 1984. *The primary source: tropical forests and our future*. New York: Norton.

Nelson, G. 1978. Ontogeny, phylogeny, paleontology and the biogenetic law. *Syst. Zool.* 27:324–345.

Nelson, G. and N. Platnick. 1981. *Systematics and Biogeography: Cladistics and Vicariance*. New York: Columbia University Press.

Nixon, K. C. 1988. *CLADOS, Version 0.9 Documentation*. Ithaca, New York.

Nixon, K. C. 1991. *CLADOS, Version 1.0 Documentation*. Ithaca, New York.

Nixon, K. C. and J. I. Davis. 1991. Polymorphic taxa, missing valves, and cladistic analysis. *Cladistics* 7. In press.

Novacek, M. J. 1986. The skull of leptictid insectivorans and the higher-level classification of eutherian mammals. *Bull. Amer. Mus. Nat. Hist.* 183:1–112.

Novacek, M. J. 1989. Higher mammalian phylogeny: the morphological-molecular synthesis. In B. Fernholm, K. Bremer, H. Jornvall, eds., *The Hierarchy of Life*. pp. 421–435. Amsterdam: Elsevier.

Novacek, M. J., A. R. Wyss, M. C. McKenna. 1988. The major groups of eutherian mammals. In M. Benton, ed., *The Phylogeny and Classification of the Tetrapods, Vol. 2: Mammals*, pp. 31–71. Oxford: Clarendon Press.

NSB (National Science Board, U.S.) 1989. Loss of biological diversity: a global crisis requiring international solution. Washington: National Science Foundation.

Patterson, C. 1981. The significance of fossils in determining evolutionary relationships. *Ann. Rev. Ecol. Syst.* 12:195–223.

Patterson, C. 1983. How does phylogeny differ from ontogeny? In B. C. Goodwin, N. Holder, C. C. Wylie, eds., *Development and Evolution*, pp. 1–31. Cambridge: Cambridge University Press.

Queiroz, de, K. 1985. The ontogenetic method for determining character state polarity and its relevance to phylogenetic systematics. *Syst. Zool.* 34:280–299.

Raup, D. M. 1979. Size of the Permo-Triassic bottleneck and its evolutionary implications. *Science* 206:217–218.

Raup, D. M. 1981. Extinction: bad genes or bad luck? *Acta Geol. Hisp.* 16(1–2):25–33.

Raup, D. M. 1984. Evolutionary radiations and extinctions. In H. D. Holland and A. F. Trandell eds., *Patterns of Change in Evolution*, pp. 5–14. Berlin: Dahlem Verlag.

Raup, D. M. 1988. Diversity crises in the geological past. In E. O. Wilson, ed., *Biodiversity*, pp. 51–57. Washington, DC: National Academy of Science Press.

Raup, D. M. and J. J. Seposki, Jr. 1984. Periodicity of extinctions in the geologic past. *Proc. Nat. Acad. Sci.* 81:801–805.

Rowe, T. 1988. Definition, diagnosis, and origin of Mammalia. *J. Vert. Paleontol.* 8:241–264.

Sanderson, M. and M. Donoghue. 1989. Patterns of variation in levels of homoplasy. *Evolution* 43:1781–1795.

Schaeffer, B., M. K. Hecht, and N. Eldredge. 1972. Phylogeny and paleontology. *Evol. Biol.* 6:31–46.

Schoch, R. M. 1986. *Phylogeny Reconstruction in Paleontology*. New York: Van Nostrand Reinhold.

Simpson, G. G. 1975. Recent advances in methods of phylogenetic inference. In W. P. Luckett and F. S. Szalay, eds., *Phylogeny of the Primates*, pp. 3–19. New York: Plenum Press.

Swofford, D. L. 1985. *PAUP. Phylogenetic Analysis Using Parsimony*. User manual version 2.4. *Illinois Nat. Hist. Surv.*

Watrous, L. E. and Q. D. Wheeler. 1981. The out-group comparison method of character analysis. *Syst. Zool.* 30:1–11.

Wheeler, Q. D. 1990a. Ontogeny and character phylogeny. *Cladistics* 6:225–268.

Wheeler, Q. D. 1990b. Insect diversity and cladistic constraints. *Ann. Entomol. Soc. Amer.* 83:1031–1047.

Wheeler, W. C. and R. L. Honeycutt. 1988. Paired sequence difference in ribosomal RNAs: evolutionary and phylogenetic implications. *Mol. Biol. Evol.* 5:90–96.

Wiley, E. O. 1981. *Phylogenetics: The Theory and Practice of Phylogenetic Systematics*. New York: John Wiley.

Wilson, E. O. 1985. The biological diversity crisis: a challenge to science. *Issues Sci. Technol.* 2(1):20–29.

Wilson, E. O. 1988. The diversity of life. In *Earth '88: Changing Geographic Perspectives.* pp. 68–81. Natl. Geogr. Soc.

Wyss, A. R., M. J. Novacek, and M. C. McKenna. 1987. Amino acid sequences versus morphological data and the interordinal relationships of mammals. *Mol. Biol. Evol.* 4:99–116.

1 : Interpreting Extinctions from the Fossil Record: Methods, Assumptions, and Case Examples Using Horses (Family Equidae)

Bruce J. MacFadden

Abstract. The interpretation of extinctions within monophyletic fossil taxa depends upon the ability to discern both the cladistic interrelationships and the temporal distribution of taxa. Although a cladogram depicts a relative sequence of speciation or branching, the temporal framework to discern origination and extinction events requires an understanding of the biostratigraphic ranges of the taxa in question. The actual age of a range can vary widely, depending upon associated geochronological data, but rarely (except during the late Pleistocene) can it be known to within several hundred thousand years for Cenozoic mammals. Older taxonomic range zone determinations (e.g., Devonian clams) can be in error by as much as tens of millions of years. Problems arise in interpretation of long-term evolutionary patterns from the fossil record because of these uncertainties and others such as sampling biases.

In many instances, the geometries of cladograms are consistent with the known temporal distributions of constituent taxa (i.e., more primitive taxa usually appear earlier in the fossil record, but this is not always the case). The most frequent exceptions are "tardy first appearances," where these datum planes, because of incomplete sampling, are younger than the actual time of origin of the taxon.

Examples are presented here from the 57-million-year history of fossil horses that illustrate the phylogenetic history and patterns and processes of speciation, origination, and extinction.

The fossil record is the fundamental basis for interpreting long-term patterns and processes of evolutionary history, including taxonomic origination and extinction. The purpose of this paper is to present and discuss the types of paleontological data that are normally used to interpret extinctions. This paper is divided into two sections. In the first, which deals with biostratigraphy and age ranges, I concentrate on examples from Cenozoic land mammals, although many of the principles presented there are also applicable to taxa that occur in other paleoenvironments (e.g., marine invertebrates) and other times in the past. In the second, I use selected case examples of phylogenetic systematics and patterns of speciation and extinction within the fossil record of horses (family Equidae).

Paleontological Units of Extinction Analysis

To understand the origin and extinction of a given taxon, paleontologists study biostratigraphic range zones, or ranges. Simply put, a range zone is the observed temporal duration of a given taxon. As discussed later, certain assumptions and caveats must be considered when using these zones. Furthermore, these limitations should be presented clearly in synthetic papers using biostratigraphic data so that they may be understood by those less familiar with the paleontological literature and practices.

In many cases the ranges of superspecific taxa are analyzed, and these can vary from the genus to family, although sometimes higher categories (e.g., order) are also used. These higher categories are often used because the specific taxonomy for a particular group is not well known. However, there is a strong feeling that the best level of biostratigraphical analysis, for any reason, whether it be evolutionary or biogeographic, is the species. Nevertheless, both the species and superspecific categories will continue to be used in evolutionary studies; and case examples of both species and genera of fossil horses are presented later. It is satisfying to note that many of the general evolutionary patterns discerned from the fossil record of superspecific taxa, and their consequent interpretations of process (e.g., Simpson 1953), are valid at various levels of the taxonomic hierarchy, including species (see Raup and Boyajian 1988). For example, in their study of early Tertiary plesiadapid primates (fig. 1.1), Maas et al. (1988) found that diversity patterns were similar at both the generic or specific levels. Simpson (1953) noted similar patterns for genera, families, and orders of fossil fishes.

When a range for a given taxon is reported to be between, say, 12.0 and 8.0 myr, what assumptions and uncertainties are involved in this statement? How confident are we that the temporal limits discerned from the fossil

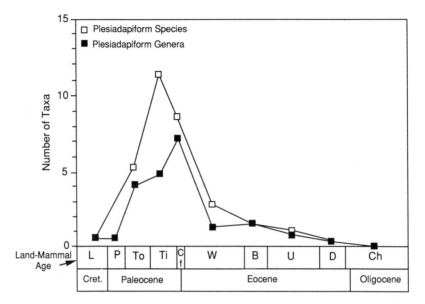

Fig. 1.1. Species and generic-level patterns of diversification (taxonomic richness) in early Tertiary plesiadapiform primates. (Modified from Maas et al. 1988: Fig. 2B.) Abbreviations along the bottom are as follows: L, Lancian; P, Puercan; To, Torrejonian; Ti, Tiffanian; Cf, Clarkforkian; W, Wasatchian; B, Bridgerian; U, Uintan; D, Duchesnean; Ch, Chadronian.

record correspond to the actual duration of that taxon? These queries are complex, but are the essence of understanding the use of biostratigraphic range zones for evolutionary analyses. For the sake of a simple illustration, and because it probably has less uncertainty than using a higher taxon, I first discuss the components of a local range zone of a single species. I then discuss wider-ranging biostratigraphic zones of species and higher categories.

Local Range Zones of Species

The fundamental unit in biostratigraphy is the local range zone of a species. I will not be concerned here with the theoretical problem of the ability of a paleontologist to recognize an extinct species that has corresponding validity to the modern biological species concept; this is discussed at length elsewhere (e.g., Cain 1954; George 1956). If a distinction needs to be made, species in paleontology can be considered "morphospecies" or "paleontospecies."

After thorough sampling of a local sedimentary basin, the observed ranges of individual species are commonly analyzed. These results are usually plotted as continuous vertical (or horizontal) lines throughout the local section where they are known to occur. In the best situations, it is assumed, rightly or wrongly, that preservational and collecting biases (e.g., for rare taxa) are negligible and that the observed range zone durations approximate the actual temporal extent of the species during its life. Like the problem of fossil species recognition, some paleontologists would cringe at such a cavalier assumption about the ability of a local stratigraphic section to preserve "faithfully" a good record of actual biological events. This is also a complex problem and is discussed at length elsewhere (Shipman 1981; articles in Behrensmeyer and Hill 1980). As the critics will be quick to point out, there are all kinds of reasons that this should not be so.

Figure 1.2 shows the characteristics of a local range zone for a hypothetical species A. It is used to illustrate some biostratigraphic principles and to discuss how its temporal duration is assessed by paleontologists. Most of the same principles also apply to local range zones of superspecific taxa (e.g., genera).

After sufficient sampling in a local basin, all fossil horizons are inventoried (see fig. 1.2). The lowest and highest stratigraphic occurrences are inferred to

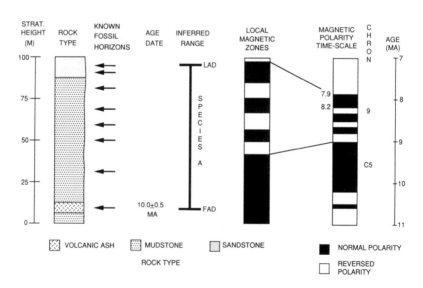

Fig. 1.2. The biostratigraphy and calibration of the range-zone of hypothetical Species A. (See discussion in text.)

represent, respectively, the origination of species A, termed a *first appearance datum* (FAD), and the extinction, or *last appearance datum* (LAD; e.g., Woodburne 1987). Although these parameters constrain the local range of species A, other data are needed to determine how old these events are. The two most popular geochronologic methods currently employed in dating terrestrial sediments are radioisotopic age determinations and magnetic polarity stratigraphy (also see fig. 1.2). Radioisotopes can be used to assess geologic time because the parent decays to the daughter element (e.g., ^{14}C to ^{14}N) at a known rate; therefore, analytical determination of these ratios in association with fossil deposits can provide an isotopic age. Magnetic polarity stratigraphy is based on the fact that the earth's magnetic field has had two dominant states in the past, i.e., there were times of "normal" (like today) and "reversed" polarities. The sequential patterns of normal and reversed polarities are global phenomena that can be used for age determination of associated fossils.

In the hypothetical example (see fig. 1.2), the lowest occurrence of species A is within a volcanic tuff deposit, which is amenable to radioisotopic dating. A radioisotopic analysis using $^{40}K-^{40}Ar$ (which is the most commonly used decay series to date Cenozoic vertebrates; see Evernden et al 1964; Woodburne 1987) yields an age of 10.0 ± 0.55 myr for this ash. (A 5% analytical error is reasonable in modern studies, although some modern instruments and new techniques can reduce the error to 1–2%). Therefore there is a probability that the actual age of the first appearance datum of species A is between 10.5 and 9.5 Ma. The local extinction (LAD) of species A, although not bracketed by a datable ash deposit, is contained within a normal polarity zone that is correlated to Chron 9 on the global Magnetic Polarity Time Scale that has limits between 7.9 and 8.2 Ma (using the time scale of Berggren et al. 1985). If the LAD occurs two thirds of the way up into the normal zone, it is tempting to extrapolate the age of this extinction to 8.0 Ma. There are, however, many reasons that this may be overextending the data, particularly because of the episodic and uncertain nature of various sedimentary processes (e.g., see McRae 1990a, 1990b). In this example, there is a change in sediment type within the normal polarity zone. The bottom one-third of this magnetic zone consists of fine-grained clays, and the upper two-thirds is sandy. As McRae (1990a, 1990b) shows, the clays could represent very slow (i.e., in the 10^5-year range) sedimentary accumulation rate; the sand could represent a short-term flood cycle occurring in a couple of days to a few months. Therefore it is generally incorrect to make a direct age extrapolation between stratigraphic position and time. If a more conservative approach is taken, then only events that occur at polarity transitions are of known age. For biostratigraphic events that occur within magnetic polarity zones, because of the frequency of reversals of the earth's magnetic field, the average uncertainty of a datum event is about 0.33 myr (Flynn et al. 1984; which happens also to be similar to the duration of the upper normal zone of Chron 9 in fig. 1.2).

Using these geochronological techniques and their associated uncertainties, the temporal duration of species A (see fig. 1.2) could be as short as 1.3 (9.5 − 8.2) myr or as long as 2.6 (10.5 − 7.9) myr. Usually, the mean ages are used in biostratigraphic analysis and the associated errors (or assumptions involved) are not mentioned. This might seem trivial, but errors can sometimes be huge for geologically older taxa, because the error is proportional to the mean age determination. A hypothetical Paleozoic clam species bracketed by radioisotopic ages (with 5% analytical error) could have a FAD of 440 ± 22 Ma and a LAD of 380 ± 19 Ma. Thus the duration of this species range, which would usually be reported in the literature as 20 Ma, could be as long as 59 (462 − 361) Ma or as short as 19 (418 − 399) Ma, an uncertainty greater than the entire time span of the Age of Mammals!

Geographically Widespread Species Ranges

The foregoing discussion centers on a hypothetical species studied from a local sedimentary basin. Often, however, local origination and extinction events are only part of the story, and one tries to determine the temporal limits for the species biostratigraphic range throughout its observed biogeographic range.

Some discussion is necessary here concerning the time span of dispersal of land mammals. Geographically widespread datum planes of mammals imply dispersal from some center of origin. If the dispersal event is very rapid (as it seems to be for some marine organisms, such as foraminifera), then the datum plane appears geologically synchronous (i.e., it cannot be discerned with modern geochronological techniques). Kurten (1957) suggested that Pleistocene mammals dispersing into North America across the Bering Land Bridge took about 10,000 years to cover the continent. Therefore if we can only resolve biostratigraphic events to within a few hundred thousand years, most datum planes are synchronous throughout the observed range of a taxon. In a few examples where dispersal seems slower, the actual difference in "arrival" of taxa can be documented, like the lagomorph *Lepus* in the Plio-Pleistocene in western North America (Opdyke et al. 1977; fig. 1.3 here), or the superspecific *Cormohipparion* complex between the New and Old Worlds (Flynn et al. 1984; fig. 1.4 here). Nevertheless, these are the exceptions to the rule, and workers who routinely believe that they can observe centers of origin and subsequent dispersal of fossil taxa, particularly species, are kidding themselves.

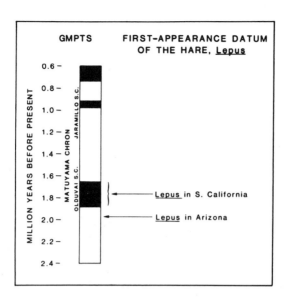

Fig. 1.3. Comparison of the calibration of the first-appearance datum (FAD) of the lagomorph *Lepus* in the San Pedro Valley of Arizona (Johnson et al. 1975) and the Anza-Borrego Desert of southern California. (Modified from Opdyke et al. 1977.)

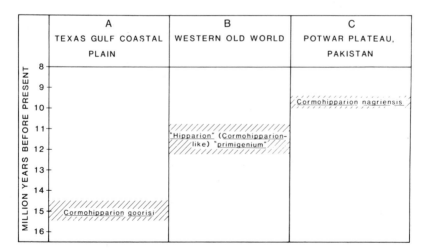

Fig. 1.4. Comparison of the first-appearance datum (FAD) of the three-toed horse *Cormohipparion* and closely related forms in North America and the Old World. The dashed zones indicate the approximate limits of temporal resolution for these biostratigraphic events. (From Flynn et al. 1984 and reproduced with permission of the University of Chicago.)

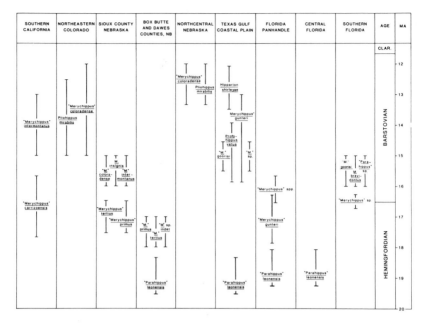

Fig. 1.5. Observed local range zones for selected Miocene horses showing composite ranges from different regions for *Parahippus leonensis* and *Merychippus gunteri*. (From Hulbert and MacFadden 1991 and reproduced with permission of the American Museum of Natural History.)

When fossil species ranges are plotted for evolutionary studies, the investigator customarily surveys the literature and determines the maximum known temporal extent of their occurrences. These data are preferred to local ranges, because the latter can have local late arrivals or extinctions. In figure 1.5, although the local range of the three-toed horse *"Parahippus" leonensis* varies within its observed paleobiogeographic range (Nebraska, Texas, and Florida), it would be quoted in the literature as having a temporal duration between about 20 and 18 Ma. Similarly, even though the ranges of *Merychippus gunteri* do not temporally overlap in the Texas Gulf Coastal Plain and Florida, the range of this species would probably be quoted as 18 to 13 Ma. The uncertainties of this statement are therefore compounded relative to the more simple example (i.e., for a biostratigraphic occurrence in a local basin, as discussed earlier).

Monographic Studies and Superspecific Range Zones

Many well-known studies of evolutionary patterns and processes of fossil groups make use of monographic compendia that list the range of taxa (usually higher categories) in time and space. For vertebrates, the systematic index of genera in Romer (1966) is commonly used, although the recent book by Carroll (1986) is both phylogenetically and geochronologically more up to date. Similarly, the exhaustive *Treatise of Invertebrate Paleontology* is often used, although now the new data base compiled by Sepkoski (1982) is both preferable and frequently cited.

Several limitations must at least be considered when reading papers that use these kinds of data. First, these compendia are almost always syntheses of many different taxonomists' decisions, and therefore the systematics can be unbalanced (e.g., splitters versus lumpers). Second, the stratigraphic ranges listed for the particular taxa are somewhat general. As a rule, the geological time scale is subdivided into roughly equal increments (e.g., early, middle, and late Miocene), and if a particular taxon is known to be of early Miocene age, it is inferred to have a duration throughout that interval. In many cases this may not be true, but it is the way that the ranges are reported and used in evolutionary studies. For example, figure 1.6 depicts a series of "observed" biostratigraphic ranges of Permo-Carboniferous fusilinids (extinct, cigar-shaped foraminifera) from the Soviet Union (original references cited in Douglass 1977). Of the 62 genera depicted in this graph, 16 (not including the questionable parts of the ranges) have both the lower and upper extents of their ranges that terminate at stage boundaries. Although many of these could be actual ranges, others probably are listed as having this duration even though they may have occurred in some narrower interval. Likewise, in the biostratigraphic ranges of North American three-toed hipparion horses (MacFadden 1984; fig. 1.7 here), *Hipparion tehonense, Nannippus ingenuum, Neohipparion eurystyle,* and *Nannippus peninsulatus* are depicted as occurring throughout the land mammal age in which they are found. It is also likely that some of these taxa did not exist for the entire period indicated, but it is impossible to constrain their ranges more narrowly at the present time.

Tardy First-Appearance Datum Planes and the Congruence Between Cladograms and Biostratigraphy

As standard practice, cladograms incorporate character states based on pre-servable morphologies of fossils and do not consider the temporal distribution of the OTUs. Subsequently, the results from the cladistic analysis are "plugged

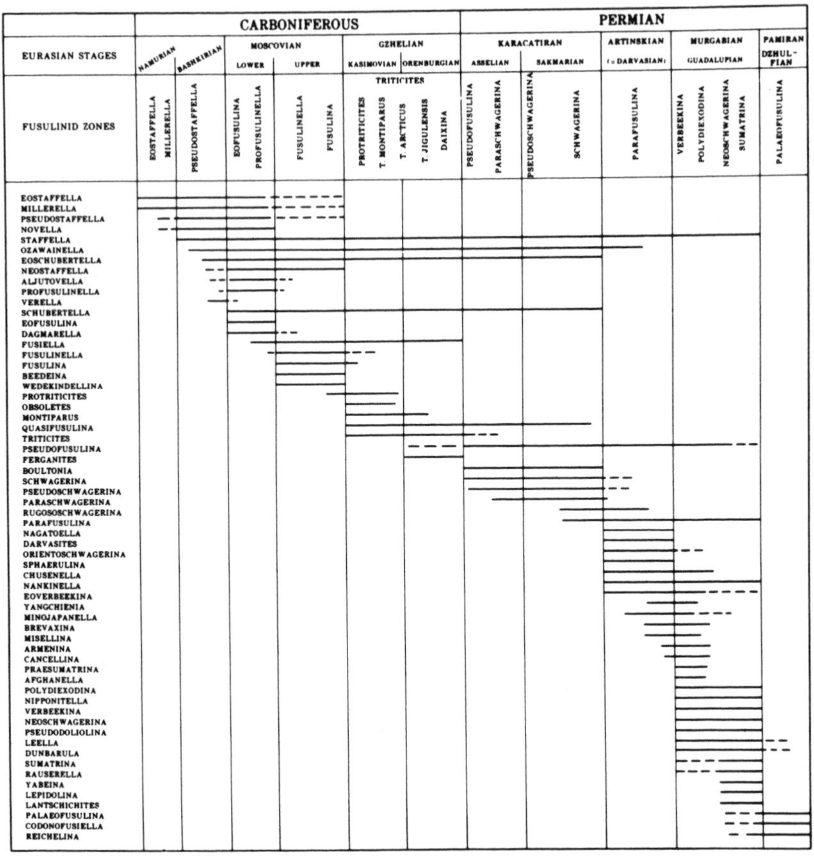

EURASIAN STAGES	CARBONIFEROUS										PERMIAN							
	NAMURIAN	BASHKIRIAN	MOSCOVIAN		GZHELIAN						KARACATIRAN		ARTINSKIAN	MURGABIAN	PAMIRAN
			LOWER	UPPER	KASIMOVIAN	ORENBURGIAN					ASSELIAN	SAKMARIAN	(= DARVASIAN)	GUADALUPIAN	DZHUL-FIAN

FUSULINID ZONES: EOSTAFFELLA, MILLERELLA, PSEUDOSTAFFELLA, EOFUSULINA, PROFUSULINELLA, FUSULINELLA, FUSULINA, PROTRITICITES, TRITICITES (T. MONTIPARUS, T. ARKTICUS, T. JIGULENSIS, DAIXINA), PSEUDOFUSULINA, PARASCHWAGERINA, PSEUDOSCHWAGERINA, SCHWAGERINA, PARAFUSULINA, VERBEEKINA, POLYDIEXODINA, NEOSCHWAGERINA, SUMATRINA, PALAEOFUSULINA

Genera (range zones):

EOSTAFFELLA
MILLERELLA
PSEUDOSTAFFELLA
NOVELLA
STAFFELLA
OZAWAINELLA
EOSCHUBERTELLA
NEOSTAFFELLA
ALJUTOVELLA
PROFUSULINELLA
VERELLA
SCHUBERTELLA
EOFUSULINA
DAGMARELLA
FUSIELLA
FUSULINELLA
FUSULINA
BEEDEINA
WEDEKINDELLINA
PROTRITICITES
OBSOLETES
MONTIPARUS
QUASIFUSULINA
TRITICITES
PSEUDOFUSULINA
FERGANITES
BOULTONIA
SCHWAGERINA
PSEUDOSCHWAGERINA
PARASCHWAGERINA
RUGOSOSCHWAGERINA
PARAFUSULINA
NAGATOELLA
DARVASITES
ORIENTOSCHWAGERINA
SPHAERULINA
CHUSENELLA
NANKINELLA
EOVERBEEKINA
YANGCHIENIA
MINOJAPANELLA
BREVAXINA
MISELLINA
ARMENINA
CANCELLINA
PRAESUMATRINA
AFGHANELLA
POLYDIEXODINA
NIPPONITELLA
VERBEEKINA
NEOSCHWAGERINA
PSEUDODOLIOLINA
LEELLA
DUNBARULA
SUMATRINA
RAUSERELLA
YABEINA
LEPIDOLINA
LANTSCHICHITES
PALAEOFUSULINA
CODONOFUSIELLA
REICHELINA

Fig. 1.6. Chart showing range zones for genera of Permo-Carboniferous fusulinids from the Soviet Union. (Taken from Douglass 1977 and reproduced with permission of Dowden, Hutchinson, and Ross.)

in" to the biostratigraphic distribution of the OTUs, usually to construct a phylogenetic tree (i.e., a chart depicting lineages of ancestors and descendants through time). But a common quip (and an early defense against the application of cladistics to the fossil record) is something like the following: What do you do if the geometry of the cladogram does not mirror that of the time distribution of taxa? The answer to this is not as worrisome as it might seem. Actually, there have been previous discussions in the literature about how cladistic geometries may be a test of the confidence in the first-appearance datum planes that we use. Novacek and Norell (1982:368) correctly pointed out "that relative sequences of splitting, like those suggested by cladograms, can be used to reveal significant inconsistencies with durations indicated by fossil taxa."

Figure 1.8 illustrates the simplest case of three taxa—A, B, and C. The character analysis indicates that A is the primitive sister taxon relative to B and C. However, the observed stratigraphic distribution of these taxa shows that A appears later in time than either B or C, and therefore using these data one might say that you cannot construct a phylogenetic tree with A as the ancestor, or more basal sister lineage, of B and C. In a situation like this, it is possible that the observed biostratigraphic range zone of A is incorrect, with a so-called tardy first appearance. Further collecting or discoveries might

Fig. 1.7. Chart showing range zones for New World species of *Hipparion, Neohipparion, Nannippus,* and *Cormohipparion.* (From MacFadden 1984 and reproduced with permission of the American Museum of Natural History.)

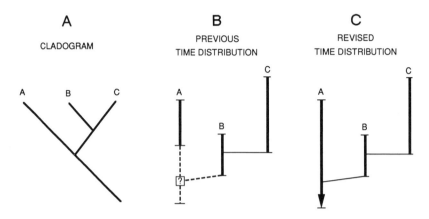

Fig. 1.8. Cladogram showing cladistic interrelationships (A) of three hypothetical taxa A, B, and C and previous (B) and current (C) range-zone distributions, the latter of which is based on additional fossil discoveries.

uncover older examples of A, thereby confirming that in a temporal sense, it could be both primitive morphologically and older than B and C.

Tardy first appearances are probably the exception to the rule, and in most cases, the primitive and derived sequence of taxa within a monophyletic group is usually consistent with the biostratigraphic distribution of its constituent OTUs. However, as pointed out by Novacek and Norell (1982), FADs should be considered minimum ages for origination-divergence times in fossil taxa.

Another anomalous situation can occur when the range zones may be "correct" for taxa but the associated age determinations are fuzzy. A good example of this is the discussion surrounding the origin of South American platyrrhine primates. A significant body of morphologic data suggests that the closest sister taxon of platyrrhines is represented by the parapithecid primates from the Oligicene Fayum locality in Egypt (Fleagle and Kay 1987). However, previous criticism (Gingerich 1980) of this linkage cite the following data (fig. 1.9): (1) The then oldest known South American primate, *Branisella,* occurs at Salla, a Deseadan locality in Bolivia that, based on the earlier calibrations of this land mammal age (Marshall et al. 1977), was thought to be about 33–35 Ma. (2) The age of the primate-bearing horizons at Fayum was thought to be about 26 Ma. Therefore it was concluded that the Fayum taxa could not be ancestral to platyrrhines (see fig. 1.9).

Subsequent data have significantly revised this interpretation. A suite of radioisotopic dates and paleomagnetic calibrations from Salla indicates that

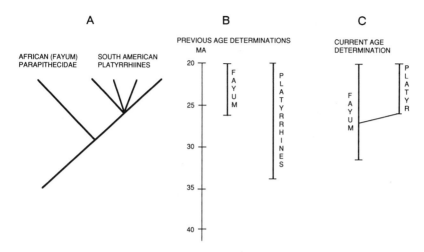

Fig. 1.9. Cladogram depicting the Parapithecidae from the Fayum of Egypt as the primitive sister taxon to South American platyrrhines (A) and previous (B) and current (C) interpretations about the age distribution of these taxa, the latter of which is based on revised radiometric calibrations at the Fayum (Fleagle et al. 1985).

the *Branisella*-bearing horizons occur at about 26 Ma, some 7–9 Ma younger than in previous interpretations (MacFadden et al. 1985; Naeser et al. 1987; MacFadden 1989). Fleagle et al. (1985) have published new ages of about 31 Ma for the important primate-bearing horizons at Fayum. Therefore the branching sequence from the cladogram is now consistent with the revised geologic calibrations of the first-appearance datum planes for the Fayum parapithecids and South American platyrrhines (see Fig. 1.9).

Fossil Horses, Cladistics, and Extinction

Perhaps more so than any other fossil group, the 57-million-year history of horses (family Equidae) has been extensively used as paleontological evidence of evolutionary patterns and processes. This use spans the literature from introductory texts in biology and geology (e.g., Stanley 1986) to advanced essays on evolution (e.g., Simpson 1944; Simpson 1953; Stanley 1979). The general biologist or paleontologist reading this literature might expect that, given the wide usage of fossil horses, very little has yet to be done with regard to the systematics of this group. This is not the case, however. Very few modern, cladistic studies have been presented in the literature until only recently. Many think that the phylogeny presented by Matthew (1930) and Stirton (1940) and reproduced in Simpson (1953) is essentially correct today. As I have noted elsewhere (MacFadden 1988), although the essential pattern of horse diversification is similar to those published decades ago, there have been many refinements and advances in our understanding of the phylogenetic relationships within this group. I attribute these advances to three factors: (1) a more precise and better-defined time scale to calibrate paleontological events; (2) the continuous influx of new and important specimens; and (3) new technology (e.g., computers) and methodologies (e.g., cladistics).

Many workers have studied rates of taxonomic evolution of fossil groups. Unfortunately, however, some have analyzed these patterns in polyphyletic or paraphyletic groups, thereby rendering their studies of dubious significance from a monophyletic, cladistic point of view (Smith and Patterson 1988). For a long time it was stated or implied in the literature that the family Equidae was paraphyletic, with the oldest known horse *Hyracotherium* (formerly *Eohippus,* now used in the vernacular eohippus) possibly giving rise to other primitive perissodactyls (related to modern-day rhinos and tapirs) as well as to horses. More recent work indicates that the Equidae (including various concepts of *Hyracotherium*) share synapomorphies that define it as a monophyletic order (e.g., MacFadden 1976; Hooker 1989). Thus their monophyly, diversity, and widespread abundance make the Equidae an excellent case study to study rates of taxonomic evolution.

There are several ways in which rates of taxonomic origination and extinction are assessed from the fossil record, as can be seen from many articles on this subject in such journals as *Paleobiology.* In this chapter I present the following measures of extinction for fossil horses and then compare these data to other selected taxonomic groups:

1. *Taxonomic survivorship.* Simpson (1953) plotted survivorship of taxonomic duration in much the same way that cohort analysis is done in population studies in ecology. This method was amplified in a very interesting article (Van Valen 1973) in which "Van Valen's law" (Raup 1975) was introduced. This law states that the probability of extinction of a taxon is constant, and this results in a linear slope of the survivorship curve. In contrast to Van Valen (1973), Raup (1975) states that this linearity only holds true at the species level. Nevertheless, survivorship curves have since been commonly employed in studies of extinction, regardless of which level of the Linnean hierarchy has been analyzed.

2. *Mean taxonomic duration.* The longevity of a species, genus, or higher category is a way to discuss taxonomic evolution as a result of turnover. Presumably, a monophyletic group with a mean species

duration of 2.5 myr has a greater turnover, and hence a greater rate of taxonomic evolution, than another with a mean species duration of 5.0 myr. This metric has been employed by such workers as Simpson (1944, 1953), Stanley (1979), and Raup (1978).

Some workers have also used the inverse relationship to express taxonomic rate of evolution. For example, Simpson (1953) stated that there were eight genera of horses within a 60-myr period, yielding a value of 0.13/myr (and also 7.5 myr/genus).

3. *Origination and extinction*. The rate at which species originate (S) balanced by the pace of extinctions (E) can be viewed as a measure to taxonomic evolution, such that $R = S - E$ (Stanley 1979). A related metric of taxonomic rate of evolution is the growth of the number of species (analog of the exponential growth curve) during an adaptive radiation (Stanley 1979), or $R = (\ln N - \ln N_0)/\Delta t$, where N_0 is the initial number of species (1 for the base of a monophyletic radiation) and N is the resulting number of species after an elapsed time Δt.

4. *Macarthurs*. Van Valen (1973) introduced the "macarthur" (ma) as a unit of extinction, where $ma = 721.3 \, (\log S_1 - \log S_2/(\Delta t \log e))$, where S_1 and S_2 are the number of surviving species at the beginning and end of time interval Δt. For small values, millimacarthurs ($ma \times 10^{-3}$) and micromacarthurs ($ma \times 10^{-6}$) also are used.

In the following section I present some recent work on fossil horses from North America and emphasize the significance of these studies to an understanding of origination and extinction. The first study, which summarizes previous results (references cited later), will examine the principal species-level diversification of grazing horses during the late Miocene. The second study presents new data on the 26 currently recognized genera of North American Cenozoic horses. As is appropriate for this volume, the cladistic interrelationships and biostratigraphic distribution will be presented; then aspects of speciation and extinction will be discussed.

Late Miocene Grazing Horses

Species-Level Patterns and Processes

Systematics. Although Stirton's (1940) phylogeny of horses attempted to depict both the species (particularly the transitional forms) and generic interrelationships, subsequent versions of this frequently cited phylogeny (e.g., Simpson 1953) only show genera. Furthermore, because many of the species-level systematics had not been determined, several genera represent horizontal-grade concepts. (Quinn [1955] was a notable exception to this practice; however, his exceedingly vertical phylogeny of horses is so radical and poorly substantiated that it has not been accepted in subsequent literature). The horizontal taxonomic scheme for fossil horses goes back to Marsh's time (1879) with an orthogenetic progression, or Matthew's (1926) concept of stages of evolution. A series of papers has dealt with the species-level cladistic interrelationships of Miocene grazing horses from North America, of which

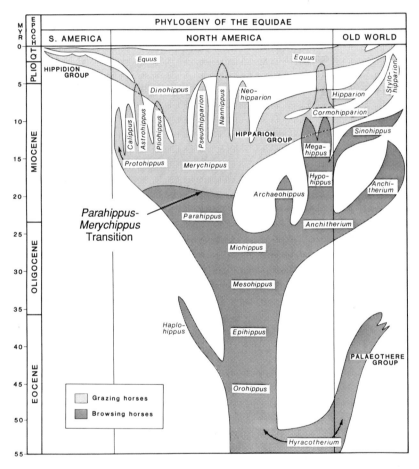

Fig. 1.10. Phylogeny of the Equidae (fossil horses) showing the systematic position of the *Parahippus–Merychippus* transition. (Modified from MacFadden and Hulbert 1988.)

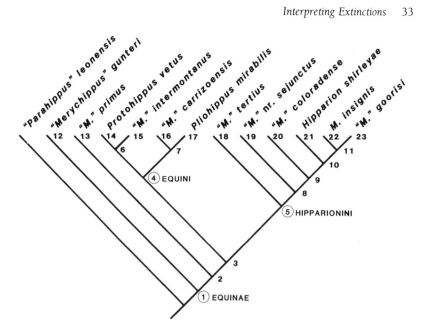

Fig. 1.11. Cladogram depicting the interrelationships of the 12 species of hypsodont horses from the Miocene of North America and the outgroup (primitive sister species) *Parahippus leonensis.* (Modified from Hulbert and MacFadden 1991; original data corroborating nodes are cited therein.)

the results from MacFadden (1984), MacFadden (1985), MacFadden and Hulbert (1988), and Hulbert and MacFadden (1991) are used here.

Parahippus-Merychippus Transition. The principal adaptive radiation of grazing, hypsodont horses occurred during the Miocene. The transition from more generalized feeders (or browsers) is traditionally depicted between the genera *Parahippus* and *Merychippus* (s. l., including *Protohippus*) in the older literature (fig. 1.10). However, two important aspects of this pattern have remained obscure: (1) the specific interrelationships and (2) whether this adaptive radiation was monophyletic. To understand these two issues better, a cladistic analysis of equid species was done for the taxonomic complex that includes *Merychippus, Protohippus,* and the Hipparion group in figure 1.10 (MacFadden and Hulbert 1988; Hulbert and MacFadden 1991).

Using PAUP (Swofford 1985), these studies analyzed 97 characters in 13 critical species during this transition. Of these, 40 proved to by phylogenetically important. A single most parsimonious cladogram involving 146 steps was produced with a consistency index of 0.58. Hulbert and MacFadden (1991) provide the details of this phylogenetic analysis; their results are summarized in figure 1.11. Of relevance here, our analysis demonstrated that relative to the primitive sister-species (outgroup) *"Parahippus" leonensis*, the remaining hypsodont horses represent a monophyletic clade, usually termed the subfamily Equinae. Within this clade there are two subclades, which are referred to as the tribes Equini and Hipparionini.

Processes of Speciation. There are certain limitations and assumptions about "reading" speciation events from the fossil record. For example, sibling species in which no morphologic change occurs in the hard parts will be impossible to detect. Also, if geographic isolation is the dominant mode of speciation, then the actual speciation event may not be preserved in the local basin where the ancestral and descendant species are observed; this is probably not a concern in stasipatric models of speciation.

Figure 1.12 illustrates two possible situations of how anagenesis versus cladogenesis can be discerned from the fossil record. On the left, a hypothetical ancestral species A gives rise directly to descendant species B. This also has been termed phyletic transformation, and the inferred last appearance datum (LAD) of species A is sometimes called a "pseudoextinction" (e.g., Stanley 1979), because it becomes a philosophical point if A actually should be considered extinct. It should also be mentioned that the recognition of the point in time when species A evolves into species B is impossible in a

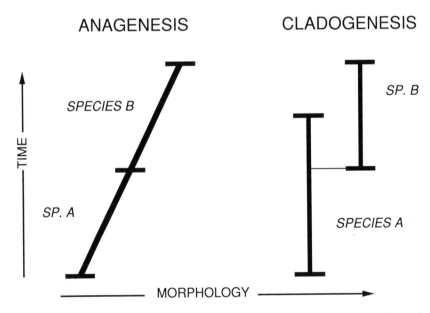

Fig. 1.12. Hypothetical ancestral Species A giving rise to Species B in the fossil record via anagenesis (left) and cladogenesis (right).

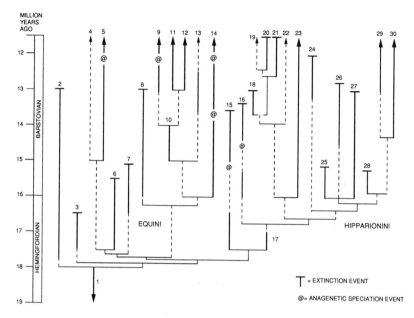

Fig. 1.13. Phylogenetic tree based on cladistic analysis of 30 species of Miocene hypsodont horses showing pattern of diversification. (Modified from MacFadden and Hulbert 1988 and Hulbert and MacFadden 1991; numbers pertain to species cited therein.)

morphologic continuum, but this mode of speciation is frequently depicted in phylogenies.

The way in which the mode of speciation is depicted in these hypothetical examples can also apply to the adaptive radiation of Miocene grazing horses in North America. If an ancestral-descendant species pair do not overlap in their observed range zones, then the speciation process is interpreted as anagenesis. On the other hand, if both ancestral and descendant species overlap for some portion of their ranges, then this is interpreted as cladogenetic speciation. Of the 30 species of equine horses that are known to occur between 18 and 12 Ma (MacFadden and Hulbert 1988; fig. 1.13 here), only six are interpreted to have originated from anagenesis, and these latter all appear later during the radiation. In a similar, partially overlapping study of the hipparions (MacFadden 1985), of the 10 species where ranges and ancestor–descendant interrelationships could be discerned, there were an equal number of anagenetic versus cladogenetic speciation events (fig. 1.14). Therefore it seems that early during this adaptive radiation the dominant mode of speciation is via cladogenesis, whereas later there is more of a

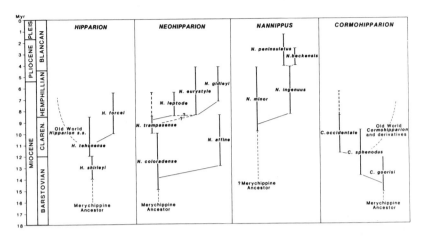

Fig. 1.14. Range zones and inferred phylogenies of the 16 species of North American hipparion species. (From MacFadden 1984, 1985; reproduced with permission of the Paleontological Society.)

balance between anagenesis and cladogenesis. This shift in mode makes sense if rapid diversification followed by equilibrium is used to explain adaptive radiations.

Species Origination and Extinction. As seen previously for horses, at the beginning of adaptive radiations the number of species increases, where origination (S) occurs much faster than extinction (E). Later during the adaptive radiations, as some equilibrium number of species diversity is reached, there is more of a steady-state balance between species origination and extinction ($S = E$). Using the exponential growth curve equation ($R = (\ln N - \ln N_0)/\Delta t$; see Stanley 1979,), for Miocene grazing horse species, Mac-Fadden and Hulbert (1988) determined that $R = 0.97$ for the beginning of the radiation between 18 and 17 Ma, whereas R dropped to 0.06 between 15 to 13 Ma (fig. 1.15). During this phase speciation (S) greatly exceeds extinction (E), and R, or net rate of species increase, is an expression of this situation. Later during the radiation, extinction and replacement occur, leaving a steady-state diversity pattern. The value of R for the beginning of the adaptive radiation of grazing horses represents high rates of taxonomic evolution in contrast to other groups, and interestingly, this is even high relative to the entire history of the family Equidae, where approximately 150 species occurred in 57 million years (table 1.1; MacFadden in preparation).

Taxonomic Durations, Survivorship Curves, and Macarthurs. MacFadden (1985) analyzed these metrics for 16 species of North American hipparion horses (figs. 1.14, 1.16). The mean species longevity was 3.33 myr. Other studies have determined mean species longevities of 6 Ma for echinoderms (Durham 1970), 1.9 Ma for Silurian graptolites (Rickards 1977), 1.2–2 Ma for Mesozoic ammonoids, and 0.2 Ma for Pleistocene mammals (Stanley 1978). For all Phaneorzoic invertebrates, Raup (1978) reported a mean species longevity of 11.1 Ma, and Simpson (1953) stated that this value is 2.5 Ma for all fossil groups.

With regard to quantification of extinction rate, Van Valen's (1973) equation (described earlier) is used here. Taking the portion of the survivorship curve (fig. 1.16B) in which most of the extinctions occur, at 2 Ma, 14 species remain (i.e., few have short durations), whereas at 5 Ma, two species remain.

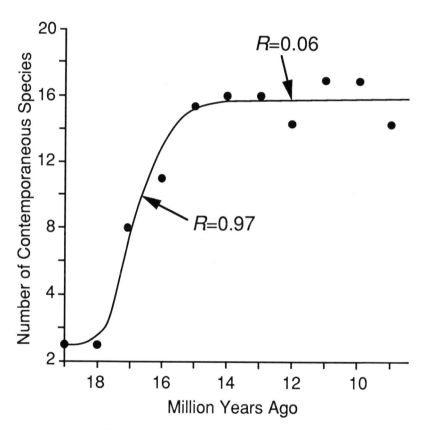

Fig. 1.15. Graph of rate of increase of species diversification for Miocene grazing horses from North America. (Modified from MacFadden and Hulbert, 1988.)

Table 1.1. Values of R (Intrinsic Rate of Increase) for Species Within Selected Fossil Clades

Taxon	Number of Species	t (Ma)	R	Reference
Miocene grazing horses 18–17 Ma	3,8	1	0.97	MacFadden and Hulbert (1988)
Miocene grazing horses 15–13 Ma	14,16	2	0.06	MacFadden and Hulbert (1988)
All fossil horses	ca. 150	57	0.09	MacFadden (in preparation)
Bovidae	115	31	0.15	Stanley (1979)
Muridae	844	19	0.35	Stanley (1979)
Cretaceous–early Tertiary eutherian mammals			0.12–0.19	Novacek and Norell (1982)
All mammals			0.22	Stanley (1979)
All bivalves			0.06	Stanley (1979)

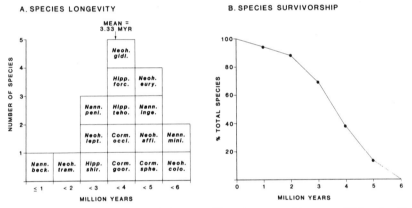

Fig. 1.16. Histogram of temporal duration (A) and survivorship curve (B) for the 16 species of North American hipparions. (From MacFadden 1985.)

This yields a species extinction rate of 468 μma. This value is high relative to those for other fossil species groups (table 1.2). One might be tempted to state that Schopf et al. (1975) and Stanley (1979) were correct that morphologically more complex groups demonstrate higher extinction rates, although the value of 334 μma for fossil primates (Novacek and Norell 1982) is lower than that calculated for hipparion horses.

Generic-Level Patterns and Processes

Systematics. MacFadden (in preparation) presents a cladistic analysis of the 26 currently recognized genera of fossil horses from North America. Although an entire reiteration of these results is beyond the intended scope of this

Table 1.2. Comparison of Rates of Taxonomic Extinction for Selected Clades of Species (in macarthurs)

Group	Rate (uma)	Reference
Diatoms	90	Van Valen (1973)
Dinoflagellates	55	Van Valen (1973)
Foraminifera	100	Van Valen (1973)
"Most groups"	5–200	Raup (1975)
Hipparion horses	468	MacFadden (1985)
Fossil primates	334	Novacek and Norell (1982)

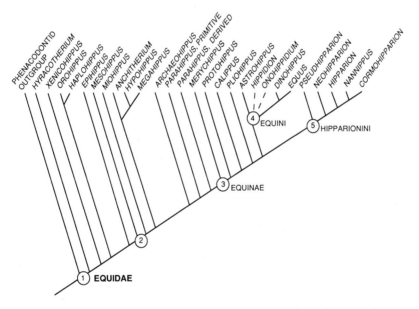

Fig. 1.17. Cladistic interrelationships of the 26 currently recognized genera of North American fossil Equidae. Characters used to justify important nodes are taken from the following references: (1) Equidae, MacFadden (1976), Hooker (1989); (2) MacFadden (1976); (3–5) Equinae, Equini, and Hipparionini, Hulbert and MacFadden (1991).

essay, the character states that justify the important nodes of the cladogram are presented in figure 1.17. The biostratigraphic distributions (from which the generic durations were taken) are depicted in figure 1.18.

From these data, the mean duration of a fossil horse genus is 8.4 myr. This is surprisingly close to Simpson's (1953) estimate of 7.5 myr and, as might be expected, lower than the mean value of 28.4 myr for all invertebrate groups (Raup 1978). Figure 1.19 depicts the generic durations and survivorship patterns for these same genera of fossil horses. The most general pattern that can be observed from these data is that, for the species level, very few genera have either very short or very long durations. Taking into account the steep portion of the survivorship curve between 4- and 12-myr duration (fig. 19B), the 26 genera of fossil horses have a rate of extinction of 161 μma. Van Valen (1973) calculates rates of 90, 55, and 100 μma for diatoms, dinoflagellates, and planktonic foraminiferans, respectively. Raup (1975) found that most fossil groups fall between 5 and 200 μma. Like the observation noted earlier for fossil species, rates of taxonomic evolution for genera may be correlated to morphologic complexity (fig. 1.20).

It should be noted that published values of taxonomic durations and extinctions may be just a rough metric of rates of evolution. All the values

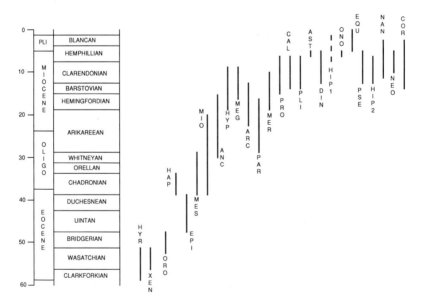

Fig. 1.18. Biostratigraphic range distributions of the 26 currently recognized genera of North American fossil Equidae.

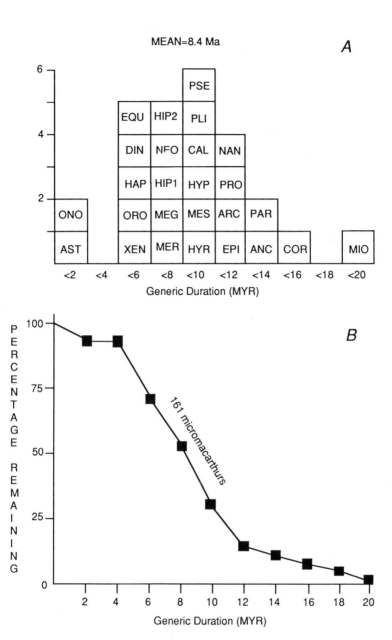

Fig. 1.19. Histogram of temporal distribution (A) and survivorship curve (B) for the 26 genera of North American fossil Equidae. The generic abbreviations are the first three letters in the genera shown in fig. 1.10 except for HIP1 = *Hippidion* and HIP2 = *Hipparion*.

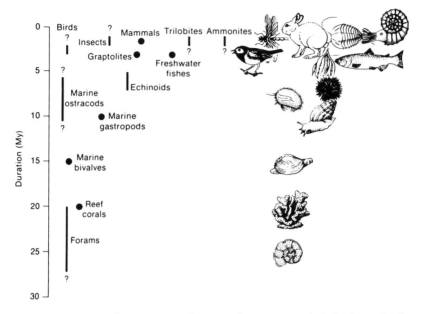

Fig. 1.20. Hierarchy of mean species duration of various animal phyla. (From Stanley 1979 and reproduced with permission of W. H. Freeman and Co.)

cited earlier may be higher than actual if one uses stratigraphic first appearance in the fossil record rather than time of phylogenetic divergence (e.g., Novacek and Norell 1982).

The Importance of Fossils to an Understanding of Extinction

Simpson (1944:144) defined taxonomic relics as "Groups once highly varied and now reduced to relatively few species." Such is the case for the fossil record of Equidae, where the extant *Equus* represents 4% of the observed generic diversity of this clade and only a similar fraction in the diversity of adaptive types. Although there are some that would argue that the fossil record is of secondary importance to systematics and phylogeny, without paleontology, there can be no study of the history of diversity and extinctions.

ACKNOWLEDGMENTS

I thank Michael J. Novacek for inviting me to present a paper in the symposium at the Hennig VIII meetings, from which this paper was written. I acknowledge the Institute for the Study of Continents (INSTOC) at Cornell University for allowing me to spend my sabbatical year there where I wrote earlier drafts of this paper. This research was partially supported by National Science Foundation grant BSR 8515003. This is University of Florida Contribution to Paleobiology number 393.

REFERENCES

Behrensmeyer, A. K. and A. P. Hill (eds.). 1980. *Fossils in the Making.* Chicago: University of Chicago Press.

Berggren, W. A., D. V. Kent, J. J. Flynn, and J. A. Van Couvering. 1985. Cenozoic geochronology. *Geol. Soc. Amer. Bull.* 96:1407–1418.

Cain, A. J. 1954. *Animal Species and Their Evolution.* London: Hutchinson's University Library.

Carroll, R. L. 1986. *Vertebrate Paleontology and Evolution.* San Francisco: W. H. Freeman.

Douglass, R. C. 1977. The development of fusilinid biostratigraphy. In E. G. Kauffman and J. E. Hazel, eds. *Concepts and Methods of Biostratigraphy,* pp. 463–481. Strouds-burg PA: Dowden, Hutchinson, and Ross.

Durham, J. W. 1970. The fossil record and the origin of the Deuterostomata. *Proc. North Amer. Paleontol. Conv.* H:1104–1132.

Evernden, J. F., D. E. Savage, G. H. Curtis, and G. T. James. 1964. Potassium–argon dates and the Cenozoic mammalian chronology of North America. *Amer. J. Sci.* 262:145–198.

Fleagle, J. G., T. M. Bown, J. D. Obradovich, and E. L. Simons. 1985. Age of earliest African anthropoids. *Science* 234:1247–1249.

Fleagle, J. G. and R. F. Kay. 1987. The phyletic position of the Parapithecidae. *J. Hum. Evol.* 16:483–532.

Flynn, J. J., B. J. MacFadden, and M. C. McKenna. 1984. Land-mammal ages, faunal heterochrony, and temporal resolution in Cenozoic terrestrial sequences. *J. Geol.* 92:687–705.

George, T. N. 1956. Biospecies, chronospecies and morphospecies. *Syst. Assoc. Publ.* (Engl.) 2:123–137.

Gingerich, P. D. 1980. Eocene Adapidae, paleobiogeography, and the origin of South American platyrrhini. In R. L. Ciochon and A. B. Chiarelli, eds., *Evolutionary Biology of New World Monkeys and Continental Drift,* pp. 123–138. New York: Plenum Press.

Hooker, J. J. 1989. Character polarities of early perissodactyls and their significance for Hyracotherium and infraordinal relationships. In D. R. Prothero and R. M. Schoch, eds., *The Evolution of Perissodactyls,* pp. 79–101. New York: Clarendon (Oxford) University Press.

Hulbert, R. C., Jr. 1989. Phylogenetic interrelationships and evolution of North American late Neogene Equidae. In D. R. Prothero and R. M. Schoch, eds., *The Evolution of Perissodactyls,* pp. 176–196. New York: Clarendon (Oxford) University Press.

Hulbert, R. C., Jr. and B. J. MacFadden. 1991. Morphological transformation and cladogenesis at the base of the adaptive radiation of Miocene hypsodont horses. *Amer. Mus. Nat. Hist. Novit.* no. 3000, 61 pp.

Johnson, N. M., N. D. Opdyke, and E. H. Lindsay. 1975. Magnetic polarity stratigraphy of Pliocene-Pleistocene terrestrial deposits and vertebrate faunas, San Pedro Valley, Arizona. *Bull. Geol. Soc. Amer.* 86:5–12.

Kennedy, W. J. 1977. Ammonite evolution. In A. Hallam, ed., *Patterns of Evolution,* pp. 251–304. Amsterdam: Elsevier.

Kurten, B. 1957. Mammal migrations, Cenozoic stratigraphy, and the age of Peking man and the australopithecines. *J. Paleontol.* 31:215–217.

Maas, M., D. W. Krause, and S. G. Strait. 1988. The decline and extinction of the Plesiadapiformes (Mammalia: ?Primates) in North America: displacement or replacement? *Paleobiology* 14:410–431.

MacFadden, B. J. 1976. Cladistic analysis of primitive equids, with notes on other perissodactyls. *Syst. Zool.* 25:1–14.

MacFadden, B. J. 1984. Systematics and phylogeny of *Hipparion, Neohipparion, Nannippus,* and *Cormohipparion* (Mammalia, Equidae) from the Miocene and Pliocene of the New World. *Amer. Mus. Nat. Hist. Bull.* 179:1–196.

MacFadden, B. J. 1985. Patterns of phylogeny and rates of evolution in fossil horses: Hipparions from the Miocene and Pliocene of North America. *Paleobiology.* 11:245–257.

MacFadden, B. J. 1988. Horses, the fossil record, and evolution—A current perspective. In M. K. Hecht, B. Wallace, and G. T. Prance, eds. *Evolutionary Biology,* Vol. 22, pp. 131–158. New York: Plenum.

MacFadden, B. J. 1989. Chronology of Cenozoic primate localities in South America. *J. Hum. Evol.* 19:7–21.

MacFadden, B. J. In preparation. Equidae. In C. M. Janis, K. Scott, and L. L. Jacobs, eds. *Tertiary Mammals of North America.* New York: Cambridge University Press.

MacFadden, B. J., K. E. Campbell, Jr., R. L. Cifelli, O. Siles, N. M. Johnson, C. W. Naeser, and P. K. Zeitler. 1985. Magnetic polarity stratigraphy and mammalian fauna of the Deseadan (late Oligocene–early Miocene) Salla Beds of northern Bolivia. *J. Geol.* 93:223–250.

MacFadden, B. J. and R. C. Hulbert, Jr. 1988. Explosive speciation at the base of the adaptive radiation of Miocene grazing horses. *Nature* 336:466–468.

McRae, L. E. 1990a. Paleomagnetic isochrons, unsteadiness, and uniformity of sedimentation in Miocene intermontane basin sediments at Salla, eastern Andean Cordillera, Bolivia. *J. Geol.* 98:479–500.

McRae, L. E. 1990b. Paleomagnetic isochrons, unsteadiness, and non-uniformity of sedimentation in Miocene fluvial strata of the Siwalik Group, northern Pakistan. *J. Geol.* 98:433–456.

Marsh, O. C. 1879. Polydactyl horses, recent and extinct. *Amer. J. Sci.* 17:449–505.

Marshall, L. G., R. Pascual, G. H. Curtis, and R. E. Drake. 1977. South American geochronology: radiometric time scale for middle to late tertiary mammal-bearing horizons in Patagonia. *Science* 195:1325–1328.

Matthew, W. D. 1926. The evolution of the horse; a record and its interpretation. *Quart. Rev. Biol.* 1:139–185.

Matthew, W. D. 1930. Pattern of evolution. *Sci. Amer.* 143:192–196.

Naeser, C. W., E. H. McKee, N. M. Johnson, and B. J. MacFadden. 1987. Confirmation of a late Oligocene–early Miocene age of the Deseadan Salla Beds of Bolivia. *J. Geol.* 95:825–828.

Novacek, M. J. and M. Norell. 1982. Fossils, phylogeny, and taxonomic rates of evolution. *Syst. Zool.* 31:366–375.

Opdyke, N. D., E. H. Lindsay, N. M. Johnson, and T. Downs. 1977 The paleomagnetism and magnetic polarity stratigraphy of the mammal-bearing section of Anza-Borrego State Park, California. *Quat. Res.* 7:316–329.

Quinn, J. H. 1955. Miocene Equidae of the Texas Gulf Coastal Plain. *Bur. Econ. Geol. Univ. Texas.* 5516:1–102.

Raup, D. M. 1975. Taxonomic survivorship curves and Van Valen's Law. *Paleobiology* 1:82–96.

Raup, D. M. 1978. Cohort analysis of generic survivorship. *Paleobiology* 4:1–15.

Raup, D. M. and G. E. Boyajian. 1988. Patterns of generic extinction in the fossil record. *Paleobiology* 14:109–125.

Rickards, R. B. 1977. Patterns of evolution in the graptolites. In A. Hallam, eds., *Patterns of Evolution,* pp. 333–358. Amsterdam: Elsevier.

Romer, A. S. 1966. *Vertebrate Paleontology.* Chicago: University of Chicago Press.

Schopf, T. J. M., D. M. Raup, S. J. Gould, and D. S. Simberloff. 1975. Genomic versus morphological rates of evolution: influence of morphological complexity. *Paleobiology* 1:63–70.

Sepkoski, J. J., Jr. 1982. A compendium of fossil marine families. *Milwaukee Public Mus. Contrib. Biol. Geol.* 51; 125 pp.

Shipman, P. 1981. *Life History of a Fossil: An Introduction to Taphonomy and Paleoecology.* Cambridge: Harvard University Press.

Simpson, G. G. 1944. *Tempo and Mode in Evolution.* New York: Columbia University Press.

Simpson, G. G. 1953. *The Major Features of Evolution.* New York: Columbia University Press.

Smith, A. B. and C. Patterson. 1988. The influence of taxonomic method on the perception of patterns of evolution. In M. K. Hecht and B. Wallace, eds., *Evolutionary Biology,* 23:127–216. New York: Plenum.

Stanley, S. M. 1978. Chronospecies' longevities, the origin of genera, and the punctuational model of evolution. *Paleobiology* 4:26–40.

Stanley, S. M. 1979. *Macroevolution: Pattern and Process.* San Francisco: W. H. Freeman.

Stanley, S. M. 1986. *Earth and Life Through Time.* San Francisco: W. H. Freeman.

Stirton, R. A. 1940. Phylogeny of North American Equidae. *Univ. Calif. Pub. Bull. Geol. Sci.* 25:165–198.

Swofford, D. L. 1985. *PAUP, Phylogenetic Analysis Using Parsimony.* Champaign, IL: Natural History Survey.

Van Valen, L. 1973. A new evolutionary law. *Evol. Theory* 1:1–30.

Woodburne, M. O., ed. 1987. *Cenozoic Mammals of North America: Geochronology and Biostratigraphy.* Berkeley: University of California Press.

2 : Fossils as Critical Data for Phylogeny

Michael J. Novacek

Abstract. Fossils have received varying emphases in studies of phylogeny. Some have argued that phylogenetic reconstruction is virtually impossible without data from the fossil record. Alternatively, fossils have been demoted to secondary ranking as phylogenetic data because they so frequently fail to preserve critical features of the once-living organisms. Recent empirical work suggests that neither extreme view is appropriate; fossils provide limited but often crucial evidence for phylogeny. One reason for the effectiveness of fossils is that the addition of extinct taxa to a data set often provides evidence of character transformation obscured by the highly specialized conditions in living taxa. Such effects have been described for higher amniote phylogeny, and, to a lesser extent, in mammals and in seed plants. Cladograms particularly vulnerable to alterations with the addition of taxa will tend to show an accumulation of traits at particular nodes, a balance of evidence that only slightly favors one topology over another, and a branching pattern that shows a high degree of resolution. Although cladograms can be affected by the addition of any taxon, fossil or extant, there is a general expectation that fossils will be particularly influential. This expectation is based on the direct correlation between ancientness and primitiveness, an argument supported by some data (e.g., higher amniotes) but not rigorously tested on a broad scale. Rather than deliberating on whether fossils are generally more or less important, it seems more productive to examine the analytical problems associated with the combined analysis of extant and partially preserved fossil taxa.

Currents of Opinion

Modern biology is built on the notion that fossils play a central role in reconstructing the evolution of life. Thus, appearing in the *New Encyclopaedia Britannica* (1986, 15th ed., 18:984) is the statement, "Paleontology occupies a key position in evolutionary studies; the fossils in the Earth's crust are objective evidence of the course taken by the living organisms in their evolutionary history, or phylogeny." This recognition for fossils was long in coming. In the sixteenth century, it was popular to imagine that fossils were stony substances corresponding in design to a network of similar patterns coursing through the living and nonliving universe (Greene 1961). Soon thereafter came the idea that some of the "stones" were actually remains of once-living creatures wiped out in the biblical deluge. It was not until the nineteenth century that fossils were distinguished as the signposts of evolutionary history. The importance of fossils has been promoted by influential scientists of the twentieth century. As G. G. Simpson (1961:67) remarked, "Modern taxonomy is evolutionary and its basis involves phylogeny, which cannot be directly observed and often must be inferred from nonpaleontological data lacking the essential time dimension." Even more emphatic is the statement that, "A phylogeny is an attempted reconstruction of the evolutionary history of a group of organisms. As such, a phylogeny can only be expected to reflect history to the extent that actual records (fossils) are available to document successive stages of evolutionary change" (Gingerich and Schoeninger 1977:483).

This reputation for fossils has failed to eradicate a historically rooted suspicion about the limitations of paleontology. Darwin (1859:249) wrote that "we have no right to expect to find in our geological formations, an infinite number of those fine transitional forms which, on our theory, have connected all the past and present species of the same group into one long and branching chain of life." In recent years this view has reemerged as a call for less dependency on fossils in phylogenetic analysis. Skeptics claimed that fossils neither are readily identifiable as actual ancestors of extant groups nor show coherent patterns of character change through time (e.g., Nelson 1978) To these critics the great potentials claimed for fossil evidence were more an apotheosis than a realistic assessment. Also noted was the incompleteness of fossils and their lack of bearing on character evidence rich with biological meaning. For instance, Goodman (1989:45) recently called for a synthesis of phylogenetic evidence under the following premise: "Such sophisticated [molecular] knowledge which weights morphological changes in terms of inferred underlying genetic changes should be particularly useful for reconstructing the phylogenetic history of fossil species since, unlike living organisms, ancient fossils are not amenable to direct genetic analysis."

Notwithstanding his evocation of *all* the pertinent evidence, Goodman relegates fossils to a comparatively low position. He maintains that fossils, because they fail to preserve genetic characters and other significant traits, are critically deficient as evidence for phylogeny. Hence fossils are to be added as a final embellishment after a tree is constructed in a first pass using molecular data, and in a second pass using morphological traits derived from "more complete" extant taxa.

It is, in fact, Goodman's (1989:45) assertion that molecular systematics epitomizes the true synthesis of cladistic methods with biologically meaningful data for powerful theories of phylogenetic reconstruction. He muses that ultimate resolution of basic problems in phylogeny lies in new molecular studies, not in further examination of fossils. However one-sided this ranking of evidence appears, Goodman's assessment does in part reflect Hennig's (1966) argument that fossils are likely to play only a secondary role in phylogeny reconstruction. Accordingly, Hennig (1981) recommended that fossils be considered only as additions to the "stem groups" leading to extant groups. (The latter are equivalent to "crown-groups" sensu Jefferies [1979] and many current workers; for a recast version of Hennig's approach, see Ax [1985].) This recommendation was consistent with the judgment that fossils seem to have little influence on the outcome of a variety of phylogenetic issues, an opinion culminating in Patterson's (1981:218) frequently quoted remark, "Instances of fossils overturning theories of relationship based on Recent organisms are very rare and may be nonexistent."

Some Test Cases

This secondary ranking of fossil evidence found inculcation in Gardiner's case study (1982) of higher amniotes. Gardiner maintained that birds were most closely related to mammals, whereas more traditional fossil-oriented classifications emphasized the remote split between mammals and other amniotes including "reptiles" and birds (fig. 2.1A). Gardiner's (1982) result positioned mammals and birds as close sister groups with progressively more remote bifurcations of crocodiles, lepidosaurs, and turtles. The revisionary cladogram (fig. 2.1B) was based on 37 shared-derived characters (synapomorphies) in the five extant amniote groups, 25 of which were anatomical or physiological characters not preservable in fossils. Løvtrup (1985) claimed to have found similar evidence for Gardiner's scheme. Acceptance of either Løvtrup's or Gardiner's cladogram would thus provide a touchstone for the previously quoted statement by Patterson; even when fossil data seem compelling, the patterns such data reveal can be swamped by contradictory evidence from extant organisms.

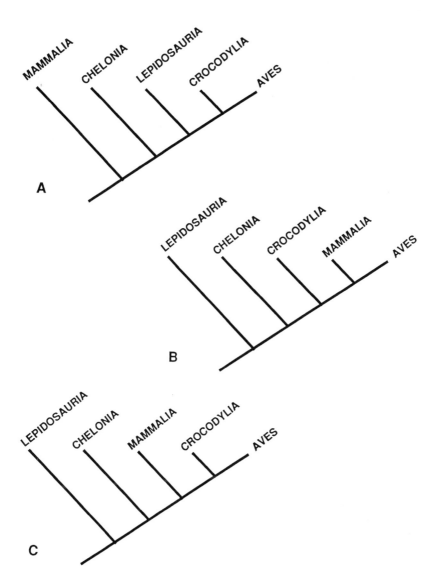

Fig. 2.1. Alternative branching schemes for higher amniote relationships. (A) Traditional arrangement, (B) Gardiner's (1982) scheme, (C) Gauthier et al.'s (1988) results using only extant taxa. See text for explanation.

The implications of Gardiner's amniote study, however, have been challenged in a reassessment by Gauthier et al. (1988). These authors made their own corrections and emendations of evidence cited by Gardiner and Løvtrup and also incorporated other character data. When Gauthier et al. (1988) confined their analysis only to extant organisms, their cladogram more closely associated crocodiles with birds, but retained the position of mammals in closer association with crocodiles and birds than was traditionally acceptable. Like Gardiner's result, turtles and lepidosaurs lie outside the mammal–crocodile–bird clade (fig. 2.1C). This vindication of the Gardiner study was disrupted when a large number of fossil taxa were added to the analysis. The resultant tree showed mammals clearly isolated from the crocodiles, birds, lepidosaurs, and turtles (fig. 2.2), the pattern conforming to the more traditional concept of amniote phylogeny (see fig. 2.1A).

One of the most diverting aspects of the Gauthier et al. (1988) analysis was the manner in which fossil taxa shifted the branching sequence. Many of the fossils had no real effect on the topology favored by Gardiner (see fig. 2.1B); that is, addition of many "reptile" fossil groups failed to sever a close connection between mammals and birds. Only the fossil members of the Synapsida (commonly known as the "mammal-like reptiles") served to isolate Mammalia from the other amniote groups (see fig. 2.2). This is because selected synapsids preserved a combination of (1) synapomorphies for special

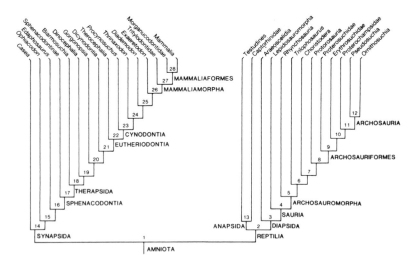

Fig. 2.2. Most parsimonious tree generated from analysis of both fossil and Recent amniote taxa. (From Gauthier et al. 1988.)

relationships with mammals and (2) primitive traits that were lost or modified in lineages related to and including crocodiles, lepidosaurs, and birds. Hence the most parsimonious cladogram (see fig. 2.2) showed a fundamental dichotomy between synapsids (including mammals) and all other amniotes (which were classified by Gauthier et al. as "Reptilia"). The authors identified the most critical range of taxa between the synapsids *Edaphosaurus* and *Exaeretodon* (see fig. 2.2). Any one of these fossil taxa by itself was sufficient to produce the tree favoring isolation of extant Mammalia from other amniote clades. One could not simply argue that this was an extraordinary case where an isolated fossil auspiciously bridged the missing link. Apparently, the fossil record preserves a diversity of taxa that strongly influence the pattern of higher amniote phylogeny.

The Gauthier et al. (1988) analysis, though perhaps the study most central to the debate over the importance of fossils, is not the only recent development on this issue. Doyle and Donoghue (1986, 1987) examined the influence of fossil data in seed plant phylogeny. Although they found less drastic changes in topology with addition or subtraction of fossils, Doyle and Donoghue did find significant effects on the robustness of certain clades. For example, Gnetales were less stable in trees limited to extant taxa than when internested with their extinct relatives in trees that combined both fossil and Recent data. Another pattern revealed in the seed plant study was the strong influence of fossils on hypotheses of character evolution. Without fossil data, the direction of certain transformations—such as derivation of linear-dichotomous leaves from pinnately compound leaves—were highly ambiguous. A more recent review by Donoghue et al. (1989) compared the major results of both the seed plant study and the amniote study and concluded that fossils play varying but often pivotal roles in phylogenetic reconstruction of living taxa. An examination of the role of fossil evidence in eutherian (placental) mammal phylogeny (Novacek 1989) demonstrated that some fossil taxa exerted minor effects on the topology of the "recent cladogram," but also greatly increased the number of alternatives among most parsimonious solutions. Rowe (1988) and Greenwald (1989) described some of the detrimental effects of including poorly represented fossil taxa in phylogenetic analysis of Mesozoic mammals. Finally, this volume demonstrates the interest the problem has attracted from a broad community of systematists.

What Are the Issues?

The amniote case (Gauthier et al. 1988) is an incisive demonstration that fossils can resolve phylogenetic events that may be out of reach of evidence derived purely from extant forms. This study, however, does not erase the possibility that further addition of extant taxa might critically overturn a phylogeny that incorporates fossils. Moreover, there is no guarantee that fossils will always have such a crucial bearing on phylogenetic questions involving extant taxa (Donoghue et al. 1989). What, then, are the important issues concerning the role of fossils as critical data in phylogeny? To begin with, it seems advisable to avoid a combat zone wherein the status of fossil evidence is under debate. I work from the premise that fossils, if available, should not be automatically relegated to some secondary purpose in any phylogenetic analysis. Rather, we need to know more about what happens when fossils are incorporated under different circumstances. Topics discussed here include effects of fossils on topology and character transformation, the problem of missing data, and the relationship between geological age and the cladistic ranking of taxa (see also Donoghue et al. 1989). In addition, this chapter covers issues related to the distinction of fossils as a special category of evidence and the alternatives for applying fossil evidence in fashioning a character-taxon matrix. Finally, I wish to review a question that has not seen much discussion—namely, how are hypotheses that incorporate fossils evaluated against either competing or coincident hypotheses based only on Recent data? The discussion will augment amniote and seed plant examples with studies of placental (eutherian) mammal phylogeny (e.g., McKenna 1987; Novacek 1989) as well as more recent work cited earlier.

Fossils: A Special Case?

Should fossils be singled out as a special problem in phylogenetic reconstruction? It is prudent to suspect phylogenies that fail to represent relevant fossil data (Donoghue et al. 1989), but suspicions apply to any study where data, whether from fossil or extant taxa, are ignored. Furthermore, attributes of critical fossils—such as an illuminating pastiche of derived and primitive states (Gauthier et al. 1989; Donoghue et al. 1989)—might be found as well in newly described extant taxa.

Why, then, should fossils be given special attention? Justification for this focus depends on the assumption that fossils, by virtue of their antiquity, are more likely to retain primitive characters than are their living relatives. As Donoghue et al. (1989) state, fossils are of particular importance because they do not continue to evolve to the present. They are more likely to preserve conditions that were subsequently radically altered over millions of years. When fossils reveal such primitive patterns, they may provide direct insight into remote splitting events that are obscured if only extant taxa are considered. Clearly, the potential for fossils to provide such insight will vary from case to case. Nonetheless, a general correspondence between age and primitiveness (see the following comments) raises our expectations for fossils in the resolution of certain phylogenetic questions. Fossil taxa do offer a temporal dimension worthy of our attention, even though extant taxa may be equally or more critical to certain phylogenetic solutions.

Methods for Incorporation of Fossil Evidence

Most arguments about the phylogenetic valence of fossils concern fossil *taxa,* focusing on whether fossil taxa should be added only after relationships of extant taxa are resolved (Hennig 1966). Or do fossil taxa demand equal status from the outset of the analysis (see Gauthier et al. 1988)? In the amniote case (Gauthier et al. 1988), five extant clades are diagnosed based on distributions of characters found in extant taxa. Simultaneously, fossil data are incorporated as twenty-four extinct amniote clades. The analysis powerfully demonstrates effects of these fossil taxa on both topology and character transformations.

Fossil evidence can, however, be applied in another, less explicit, fashion. Such data can be used to enrich the character description for any OTU (terminal taxon) that includes both fossil and extant taxa. Consider, for example, the recognition of the mammalian order Carnivora as a terminal taxon. All living Carnivora lack a clavicle (or collarbone). Restriction of the concept Carnivora to living taxa would establish the absence (inferred loss) of the clavicle as an ancestral condition and a synapomorphy for this group. Inclusion of fossil evidence would modify this characterization, because certain fossils (e.g., Early Tertiary miacids) assignable to Carnivora on the basis of other characters have a clavicle and this element is present in the probable close relatives of Carnivora (see Novacek and Wyss 1986). The absence (inferred loss) of the clavicle can no longer be used as a diagnostic synapomorphy for Carnivora as a whole. In this way, fossil taxa are not explicitly included as terminal taxa, but the characters they bear have an important effect on construction of the taxon-character matrix. This implicit approach may lead to a reordering of character transformations that mirrors those derived from the use of fossils as distinct terminal taxa (for the latter, see Donoghue et al. 1989).

The use of fossils merely to alter descriptions of selected terminal taxa is commonly, if tacitly, applied in many studies. It is logical that such data cannot be ignored in developing a diagnosis for a group that includes both fossil and extant members. A problem arises, however, with this application;

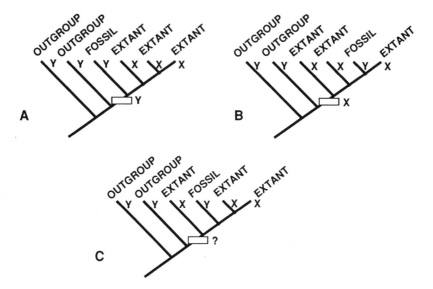

Fig. 2.3. Use of fossil evidence in hypothetical examples. Optimization of character state changes X or Y where (A) a fossil is the sister taxon of the three Recent groups, (B) a fossil is nested at a lower level within the Recent group, (C) a fossil occupies a position intermediate to those of A or B. See text.

it may be too anecdotal to ensure a reasonable estimate of ancestral condition for the group in question. In fact, if fossil taxa are not explicitly recognized in the analysis, the use of "fossil characters" may even misrepresent the ancestral condition. In a simple example shown in figure 2.3, a group has two expressions for a character (X or Y). All three living members of the group show condition X, whereas the only fossil member of the group shows character Y. Two outgroups also show character Y. In figure 2.3A, optimization of this character at the ancestral node for the group is straightforward. Because the fossil is the sister taxon of all three Recent taxa, its conformity to conditions seen in outgroups identifies Y as the ancestral condition of the group. In figure 2.3B the fossil taxon, based on characters independent of {X, Y}, is nested at a lower level within the group. Here the two earliest branches of the group are represented by extant taxa with condition X. Global parsimony (accounting for both ingroup and outgroup variation) favors the origin of character state X at the ancestral node of the group. If the fossil taxon is shifted one node higher than in figure 2.3B, the ancestral condition for the group is ambiguous (fig. 2.3C). The danger of applying fossil evidence purely as character information is thus apparent. Only when the position of the taxon is established will its bearing on the diagnosis of the ancestral node for

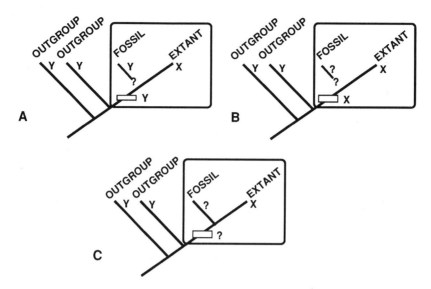

Fig. 2.4. Implicit versus explicit use of fossils for diagnoses of groups. (A) Fossil with character state Y forces rediagnosis of a group (outlined by a rectangle) that also includes an extant lineage. (B) Fossil lacking either character state X or Y does not affect diagnosis based on extant taxa. (C) Fossil used explicitly as a separate clade indicates ambiguity for trait at the ancestral node. See text.

its group be precisely determined. It might be countered that fossil data used to modify diagnoses are safely assumed to represent basal taxa (fig. 2.3A). In reality, many studies involve terminal taxa for which the internal cladistic structure of both fossil and extant members is poorly known.

Another dimension to this problem concerns the use of fossils with missing data. An implicit use of an incomplete fossil for character diagnosis can obscure meaningful patterns even if the relationships of the fossil taxon are reasonably well established. In figure 2.4A the fossil taxon, by virtue of its cladistic position, forces a rediagnosis of a group containing an extant lineage with character state X. The fossil, like the outgroups, is known to possess state Y. In figure 2.4B, the fossil is missing the character whose variable states exist in both the extant ingroup taxon and the two outgroups. Although the fossil taxon is, on other counts, a member of a group that includes the extant taxon with state X, this taxon will have no relevance to evaluation shown in figure 2.4B. In other words, with respect to this character, the fossil does not exist. The assignment at the ancestral node accordingly would be a novelty (state X) exclusive to the extant taxon. This is a problematic result. If the fossil from the outset were treated as a separate clade, optimization of the character states could identify the ancestral node as ambiguous for the character in question (fig. 2.4C). Under this modus operandi, the incomplete nature of the data with reference to the character would be explicitly recognized, and this recognition could bear on both character optimization and identity of the ancestral condition for a taxon.

Given the obvious drawbacks to the use of fossils in this implicit manner, one might ask if such an approach is ever justifiable. The defense of this application hinges on some practical considerations. In the amniote case, parsimony analysis of 24 extinct taxa and five extant clades proved enlightening (Gauthier et al. 1989). There is, however, no guarantee that such a large number of fossil terminal taxa will always be so informative. Taxa having a significant percentage of ambiguous character states can represent unwieldly and irresolvable data. Addition of even a few incompletely preserved fossils can explode fairly coherent patterns based on more complete taxa (Rowe 1988; Greenwald 1989; Novacek 1989). In fact, noise in the form of homoplasy will increase with increasing numbers of taxa (Sanderson and Donoghue 1989) as long as some contradictions in character distributions exist. At some level, a diagnosis for a taxon is simply an *estimate* of the ancestral condition that best represents the diversity within the group. Even the diagnoses of some of the fossil genera used as terminal taxa in the amniote study (Gauthier et al. 1988) summarize the variation of traits among their member species.

Such estimates of general conditions for taxa represent compromise solutions to the problem of polymorphism. Polymorphism is a reality whether the groups analyzed have fossil members or not, and it confounds efforts to diagnose OTUs. Polymorphism may be accounted for in available algorithms (such as one included in Swofford's PAUP version 3), but these algorithms will not erase serious ambiguities that arise with polymorphism. Hence estimated group diagnoses are simply a means of proceeding with the analysis in the face of prodigious character and taxic diversity. Such estimates are not exclusive to higher taxa (e.g., orders or phyla) but are integral to statements on the ancestral condition for lower-level taxa. The effectiveness of these estimates will vary according to the degree of polymorphism, the degree of internal cladistic structure, and so forth. Given a choice, one might prefer to include any fossil evidence bearing on an estimate for the ancestral condition of a group, even if the relationships of the fossil taxon as well as many of the extant taxa assigned to the group are not rigorously identified.

Effects: Examples from Amniotes, Seed Plants, and Mammals

Gauthier et al. (1988) and Donoghue et al. (1989) have identified sources of alterations of topology and character transformation with the addition of taxa.

Their discussion centers on fossils as a particular category of added data under the assumptions mentioned earlier, but the conditions specified could allow any added taxon, whether fossil or extant, to promote topological change. Statements derived from the amniote and seed plant examples are considered here in the light of a third case, the higher phylogeny of eutherian mammals (Novacek 1989). This last case demonstrates a less decisive impact of fossil taxa on topology than the amniote case. However, the generalizations offered by Donoghue et al. (1989) are consistent with results of the eutherian mammal study.

Several conditions (following Donoghue et al. 1989) seem to increase the likelihood that the addition of taxa (whether fossils or extant forms) will promote major changes in topology.

1. *Large "gaps," where clades are distinguished by nodes with numerous apomorphies, occur in the cladogram.* It seems improbable that a densely supported node always represents massive character change (saltation) coincident with one splitting event. Such enriched nodes alternatively suggest that "intermediate taxa" have not been discovered. Under this assumption, we would expect that the addition of taxa will provide critical insight on relationships of major groups. Higher amniote groups offer an excellent example of this point. The modern amniote classes (birds, mammals, lepidosaurs, crocodylians, and turtles) can be diagnosed by impressively large sets of characters (Gauthier et al. 1988). Although these features are useful in inscribing a bold case of the monophyly of each taxon, they provide limited and even potentially misleading information on higher amniote relationships. The addition of fossil amniotes produces a series of intermediate steps that are less removed from the base of the amniote cladogram. Some autapomorphies for extant taxa are now distributed as synapomorphies at a higher cladistic level. New characters highly relevant to basal branching are also added to the data set. This is because certain characters found in fossil taxa have been either lost or so radically transformed in extant groups that the latter are essentially noninformative for those characters (see Gauthier et al. 1988 and following comments).

2. *A balance of evidence favors alternative relationships.* Added taxa tend to swap critical branches if evidence for alternative groupings in the original tree is not decisively one-sided. In other words, if controversial aspects of the most parsimonious cladogram are supported by limited character evidence, the addition of taxa is more likely to change the position of the most tenuous clades.

3. *The added taxa have character combinations that introduce conflict, usually by functioning as "plesiomorphic sister taxa."* In the most extreme case, such critical taxa share at least one derived trait with at least one of the terminal

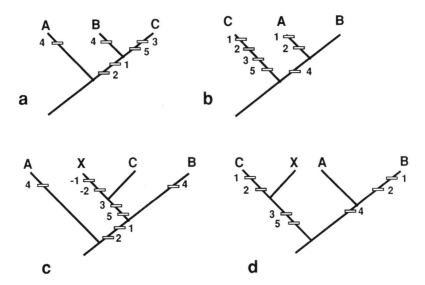

Fig. 2.5. Effects on topology with the addition of taxon X. See text.

taxa but possess the ancestral condition for every other trait. They thus fulfill the role of "plesiomorphic sister taxa."

This argument is better explained using a hypothetical example like the one provided by Donoghue et al. (1989, figs. 7 and 8). (Figures 7b and d in Donoghue et al. are problematic because one grouping, ABDE, is shown without a supporting character. Instead, taxon A should be extended to the base of the cladogram as an element of an unresolved trichotomy. Otherwise, the fully dichotomous solutions in figs. 7b and 7d would be at least two steps longer than their designated length of seven steps. To correct for this ambiguity, a simplified example is presented here in fig. 2.5) The most parsimonious cladogram for three taxa (fig. 2.5a) favors the grouping BC by one step over grouping AB (Fig. 2.5b). Addition of taxon X on the topology of Figure 2.5a introduces a character conflict because this taxon shares with C the derived traits 3 and 5 but lacks all other characters in the data set. The retention of group BC with the inclusion of taxon X requires additional steps represented by reversal of characters (1, 2) supporting BC (fig. 2.5c). As Donoghue et al. (1989) note, taxon X has the effect of "neutralizing" characters that formerly were unequivocal in supporting B and C (e.g., fig. 2.5a). A shorter cladogram is produced by moving CX to a basal position as the sister group of AB (fig. 2.5d). An important point raised by Donoghue et al. is that the neutralizing effect of taxon X is not in itself sufficient to force the outcome in one direction. There must also be prior evidence for a grouping that excludes CX. In figure 2.5, this evidence is character 4, whose distribution conflicts with the distributions of characters (1, 2) supporting BC. This is another way of stating that there must be a reasonable balance of evidence for alternatives (point 2) prior to the addition of taxa.

It might be argued that the hypothetical example in figure 2.5 negates the first condition noted earlier. The data set does not show large "gaps" in the form of abundant autapomorphies for each terminal taxon. The example is, on the contrary, the simplest case that illustrates the bridging of such "chasms of autapomorphy" with transitional taxa. Character 3 is initially known only as an autapomorphy for taxon C. Subsequent "discovery" of a taxon (X) that shares character 3 and lacks all other characters has a critical bearing on relationships of taxon C. Taxon X thus serves as a "plesiomorphic sister taxon" in much the way that any one of several synapsid fossil groups serve to isolate mammals from the other amniote clades (Gauthier et al. 1988). The array of synapsid groups form a discrete series of steps between the basal node of Amniota and the Mammalia (sensu Gauthier et al. 1988; see fig. 2.2 herein). Indeed, marked alterations to topologies are most expected when new taxa are arranged in a pectinate (comblike) cladogram (Donoghue et al. 1989).

A consideration of pectinate or alternative topologies for cladograms can be aimed at a more general level. The example in figure 2.5 shows alternatives in the form of discrete sets of dichotomies, such as are encountered in the case of higher amniotes (see fig. 2.1) or higher seed plants (fig. 2.6). Very often, however, data sets do not produce alternative cladograms that are so highly resolved. Many problems, such as those concerning major groups of birds, mammals, insects, rodents, and other important taxa, have poorly resolved sectors, or polytomies. In cladistic analyses these polytomies often represent the consensus solutions for all the equally parsimonious conflicting arrangements of clades. When character conflicts become so rampant, new taxa, even those with the optimal combinations of characters noted earlier, will naturally have less potential to alter topologies or resolve polytomies. This argument can be stated in simple form and added to those defined by Donoghue et al.

4. *Prior to addition of taxa, alternative cladograms show highly resolved dichotomies among their basal branches.* Note that this statement does not detract from the point that the initial alternatives show some balance of support—it only refers to the degree of resolution in the alternatives, not to the relative strength of one alternative over another. Put another way, this condition does not preclude conflicting branching schemes, it only maintains that those conflicts are limited to a discrete set of problems. In the amniote

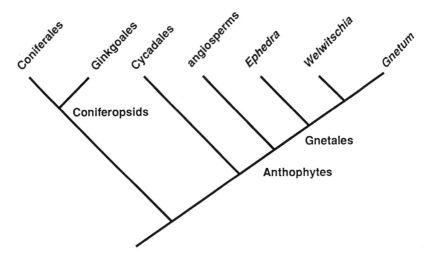

Fig. 2.6. Representative most parsimonious tree for several extant seed plant taxa based on analysis of Doyle and Donoghue (1986). This tree shows the same topology whether Recent or combined (Recent plus fossil) data are used.

case, initial debate centered on whether birds were more closely related to mammals than to crocodilians. Even in its expanded form, the problem dealt with alternative highly resolved branching schemes among five extant taxa (see fig. 2.1). Contrast this with placental mammals, where each one of the eight shortest trees derived from the Recent data set (tables 2.1 and 2.2) has polytomies involving at least five major clades (fig. 2.7). It is sobering to consider the limited value of added taxa (fossil or otherwise) in questions concerning major biological groups whose "fan-shaped" interrelationships show poor resolution.

The preceding four generalizations can be applied to the contrast between results of the amniote case and other studies. Doyle and Donoghue (1986) showed that 13 fossil taxa did not have a marked influence on relationships among seven extant seed plant groups. The primary effect of the fossils on topology was simply to increase the support for certain groupings established by examination of extant taxa. For example, Gnetales was nested within Anthophytes in the most parsimonious tree of both Recent and combined (fossil and Recent) data sets (fig. 2.6). Exclusive use of Recent data produced a tree one step longer with Gnetales nested within coniferopsids. The combined data set produced this arrangement in a tree a minimum of four steps longer than the most parsimonious tree. Donoghue et al. (1989:442) remarked, "It appears then that fossil seed plants somehow solidify relation-

Table 2.1. Character States for 88 Morphologic Characters for 20 Extant and 7 Fossil Mammalian Taxa

Characters

```
0000000001111111111222222222233333333334444444444555555555566666666667777777777888888888
1234567890123456789012345678901234567890123456789012345678901234567890123456789012345678

0000000000000000000000000000000000000000000000000000000000000000000000000000000000000000  OUTGRUP
0000000000000000000000000000100000000000000000000000000000000000000000000000000000000900  MONOTRM
1111111111000000000000000000000000000100000000000000000000000000000000000000000000000000  METATHR
1111111110011111111111111011110000000000000000000000000000000000000000000000000000000000  EDENTAT
1111110000111111111111111011110000010000000000000000000000000000000000000000000099009     PHOLIDT
1111110000111111110000001100000001000000000000000000000000000000000000000000000100        CARNIVR
1111110000111111110001000100000000000000000000000000000000000000000000000000000000111      TUBULDN
1111110000111111110000001000000010000000000000000000000000000000000000000000000011111      INSECTV
1111110000111111110000001110000000000000000000000000000000000000000000000000000011110000   PRIMATE
1111110000111111110000001110000000000000000000000000000000000000000000000000000011100000   SCANDEN
1111110000111111110000001111111000000000000000000000000000000000000000000000000000000009   DERMOPT
1111110000111111111111111111111100000000000000000000000000000000000000000000000000000000   CHIROPT
1111110000111111110000001000000001000000000000000000000000000000000000000000000000001000   MACROSC
1111110000111111111111111110000001000000000000000000000000000000000000000000000000001000   LAGOMOR
1111110000111111110000001111111100000001000000000000000000000000000000000000000000001000   RODENTS
1111110000111111110000001000000011000000000000000000000000000000000000000000000000000000   ARTIODC
1111110000111111110000001000000011000000000000000191010000100000000000000000000000000000   CETACEA
1111110000111111110000001000000011000000000000000001111110000000111110000000000000000000   PERISDC
1111110000111111110000001000000011000000000000000001111111000001000000000000000000001010   HYRACOI
1111110000111111110010001000000011000000000000000001111111110001111110001000011000009000   SIRENIA
9999199900009999991999910909099199000090090000090900000009900099909090909090900009000     PROBOSC
9999199900009999991999910909099199000090090000099000990099999909090909090900090900011111   DESMOST
99991999000099999909999109090991990000900900000900009099009990999090909090909000900900100  LEPTICD
9999199900009999991999910909099199000090090000091900090099909090909090900009000            MICROSY
9999199900009999991999910909099199019000090000010999009999099909090909090901109900         PLESIAD
9999199900009999991999919999909991999990099990099999999099990990990990990990990990         ANAGALD
9999199900009999991999919990909991999909999990999999099909990990990990990990990990990       KENNALS
9999199900009999991999919990909991999909999990999999090999999909990990990990990990          ASIORYT
```

NOTE: 0 = primitive condition; 1 = derived condition described in Table 2.1; 9 = uninformative condition due to preservational loss or marked transformation. For acronyms of groups see figures 2.7 and 2.12. Numbers underlined indicate derived conditions that would change (1>0) if characters of fossil members of the terminal taxon were considered.

Table 2.2. Characters for Higher-Level Mammalian Groupings*

---

1. Anal and urogenital openings separate in adults
2. Eggshell absent
3. Unilaminar blastocyst stage present
4. Microlecithal egg with holoblastic cleavage
5. Cochlea spiral, with at least one turn
6. Supraspinous fossa of scapula present
7. Mammary glands with teats
8. Development of chorioallantoic membrane suppressed
9. Pseudovaginal canal present
10. Ossified contribution to auditory bulla by alisphenoid
11. Last premolar is only cheek tooth replaced
12. Dental enamel with distinct tubercles
13. Sphenorbital fissures confluent below dorsoventrally compressed orbitosphenoid
14. Modified trophoblast and inner cell mass
15. Definitive chorioallantoic placenta present
16. Prolonged intrauterine gestation
17. Ureters pass lateral to derivatives of Müllerian ducts
18. Corpus callosum connects cerebral hemispheres
19. Epipubic bones and "marsupium" absent
20. Upper molars (primitively) with narrow stylar shelves
21. Fusion of Müllerian ducts into median vagina
22. Shell membrane absent
23. Seminal vesicle present
24. Retinal cones of eye simple, lacking oil droplets
25. Optic foramen distinctly separated from sphenorbital fissure
26. Frontal with pronounced ventral expansion in orbital wall, contacts ventrally confined palatine
27. Alisphenoid ventrally confined in orbital wall, not in broad contact dorsally with parietal
28. Lack of interparietal ossification in embryo and adult
29. Tooth development suppressed, anterior teeth reduced or lost, and enamel poorly developed or absent
30. Rectus thoracis muscle present
31. Subarcuate fossa very shallow or absent
32. Mastoid exposure in ventral basicranium and occipital region reduced or absent (see character 60)
33. Stapes bicurate, stirrup-shaped with large stapedial foramen
34. Foramen ovale completely enclosed with alisphenoid, not notched into posterior margin of this element
35. Pendulous penis suspended by reduced sheath between genital pouch and abdomen
36. Sustentacular facet of astragalus in distinct medial contact with distal astragalar facet
37. Ribs flattened especially near vertebral ends
38. Marked elongation of forelimbs
39. Patagium (flight membrane) continuously attached between digits of manus
40. Occipitopollicalis and humeropatagialis muscles present in flight membrane
41. Concomitant proximal and distal reduction of the ulna

42. Fenestra rotundum of cochlea faces directly posteriorly
43. Subarcuate fossa greatly expanded and dorsal semicircular canal clearly separated from endocranial wall of squamosal
44. Embryonic disc oriented toward mesometrial pole of uterus at time of implantation
45. Canals in external alisphenoid for transmission of masseteric and buccinator branches of maxillary nerve
46. Paired incisive foramina well developed, elongated, open posteriorly on palate, but nasals not recessed
47. Posterodorsal process of the premaxilla well developed, contacts frontal, but no backward extension of nasal opening or nasal elements (see character 68)
48. Contact between maxilla and frontal confined by premaxilla above and lacrimal below
49. Infraorbital canal reduced to a foramen within anterior root of zygoma
50. Premaxilla and maxilla of roughly equal exposure on palate
51. Glenoid fossa set well dorsally of basicranium, concave anteroposteriorly oriented trough lacking a postglenoid process
52. Postglenoid foramen shifted, opens in lateral eminence of squamosal
53. Two or three closely spaced hypoglossal foramina present
54. Upper and lower I1 lost, large upper and lower ever-growing deciduous I2 and lower I3 lost
55. Fetal development involves hemochorial placenta differentiating opposite attachment of blastocyst at abembryonic pole
56. Allantoic diverticulum small, rudimentary
57. Clavicle greatly reduced or absent
58. Bunodont crowns on molars, well-developed hypocones, lower molars show swelling around base of metaconid and elongation of talonids on m3
59. Inflated, pneumatic tegmen tympani (auditory roof) present
60. Amastoidy; mastoid process concealed in ventrolateral cranium by expansion and posterior overlap of squamosal on the occiput
61. Jugal extends posteriorly as a prominent ventral crest to anterolateral border of the glenoid fossa
62. Carpals dorsoventrally compressed and serially arranged, lunar–unciform contact very weak or absent
63. Fetal membrane development comprises zonary placenta in association with a free reduced yolk sac (in later stages) and an enlarged, saccular allantoic vesicle
64. Bilophodont cheek teeth with tendency to form additional lobe on posterior part of cingulum
65. Forward displacement of orbits
66. Infraorbital canal very short, ventrally situated, floored by a narrow bridge of the anterior zygoma
67. Zygomatic process of the squamosal robust, strongly produced dorsally and laterally
68. Premaxilla with very strong posterodorsal process extending around retracted nasals, nearly contacts frontals
69. Extensive sacroinnominate fusion and complete or nearly complete closure of ischiatic notch to form sacroischiatic foramen
70. Inferior ramus of the stapedial artery and inferior petrosal nerve run dorsal to tegmen tympani
71. Pancreas diffuse

Table 2.2. Characters for Higher-Level Mammalian Groupings (*Continued*)

72. Musculus styloglossus bifurcate
73. Cochlear canaliculus absent
74. Tridactyl pes
75. Large eustachian sac in tympanic region
76. Extracranial course of the internal carotid artery
77. Tuber maxillaris present
78. Lacrimal with distinct process
79. Penial glandular fossa present
80. Complete postorbital bar
81. Petrosal-derived osseous canals around intratympanic portions of facial nerve and stapedial artery
82. Anterior carotid foramen in basisphenoid converted to a long tube
83. Tegmen tympani expanded anterolaterally to roof epitympanic recess; tegmen with an epitympanic crest within which the stapedial artery runs
84. Broad maxillary–frontal contact in facial region between nasals (above) and lacrimal (below)
85. Lacrimal confined to orbit and orbital rim, lacking a facial process
86. Sphenopalatine and dorsal palatine foramen in a common recess
87. Glaserian fissure distinct, elongate trough on anterolateral wall of the tympanic cavity
88. Large mastoid tubercle (incorporates tympanohyal) nearly reaches lateral edge of promontorium cochleae

* For discussions of characters, see Novacek (1989) and Novacek and Wyss (1986).

ships among extant groups." (Groups with plesiomorphic sister taxa can, however, be highly unstable; see the following comments.

Why was the effect of fossil seed plants on topologies so subtle? After all, some of the conditions noted earlier—such as the high number of autapomorphies in selected terminal taxa (condition 1) and the number of conflicting characters suggesting closely competitive alternatives (condition 2)—are met by the case involving extant seed plants. Donoghue et al. (1989) suggested that added taxa (in this case, fossils) had a minimal effect because, rather than introducing character conflict, they simply strengthened the favored hypothesis based exclusively on extant taxa (in other words, condition 3 is not strictly met). Fossils more emphatically supported the relationship between Gnetales and angiosperms preserved in the most parsimonious "Recent tree." Additional taxa can be valued for their potential to corroborate as well as to refute hypotheses of relationships. Nonetheless, the seed plant example fails to promote the argument that fossils contribute greatly to revising topological patterns derived from study of extant forms.

A second contrast between the seed plant case and the amniote case noted by Donoghue et al. (1989) was the observation that none of the extant seed plant groups showed the high number of reversals found in mammals in the analysis of extant amniotes. They argued that reversals in the latter lead one to suspect that the group is artificially nested within the Aves–Crocodilia–Chelonia assemblage and that additional taxa could shift mammals to a more basal position. This shift to a higher-level node results in replacement of reversals by convergences. The authors thus caution that investigators should be wary when they come upon taxa, such as mammals in the amniote data set, that have numerous apparent reversals. I find problematic, however, the suggestion that marked reversal a priori may indicate an artifact of the data. The reasoning (Donoghue et al. 1989) that convergences may be more biologically acceptable—and thus more likely—than reversal is short of compelling. Many reversals of identified character states can be readily associated with the plasticity of ontogenies. Furthermore, it is not at all apparent why trees that accommodate numerous reversals are more vulnerable to topological shifts than are data that show homoplasy largely as convergences.

The study of eutherian (placental) mammals shows effects of fossils that are much less marked than those in the amniote case. Unlike the seed plant example, however, addition of fossil information did promote one important topological change that bears on the controversial position of hyraxes. This was not evident in an earlier assessment of the problem (Novacek 1989) for a simple procedural reason. In that study, I added seven extinct clades to a revised data set comprising 20 orders of extant mammals and 88 morphologic characters. This addition resulted mainly in proliferation of equally parsimonious solutions because extinct taxa were poorly represented by characters and their known traits provided little evidence of special relationships with other taxa. The analysis, however, was not strictly a test of the effects of fossils on a Recent data set. It was initiated with a series of extant terminal taxa (orders) some of whose diagnoses, following the approach outlined earlier, were estimates of ancestral conditions based on both fossil and extant members. The fossil "ingroup data" can be removed in order to examine their underlying influence on topology. This is another way of discovering how many "false synapomorphies" are indicated when only character data of extant taxa are considered.

Figure 2.7 shows eight shortest trees derived using HENNIG 86 (Farris 1988) and PAUP (Swofford 1985) from an 88-character matrix for 20 extant mammalian orders where fossil ingroup data have been removed. This subtraction forced a change in state for a total of only seven characters distributed among four mammalian orders (table 2.1). Fossils thus contributed minimally to diagnoses of orders based only on extant members. Nonetheless, modifications are sufficient to promote two changes in topology. Figure 2.8 is a consensus tree that preserves consistent elements of the trees shown in figure

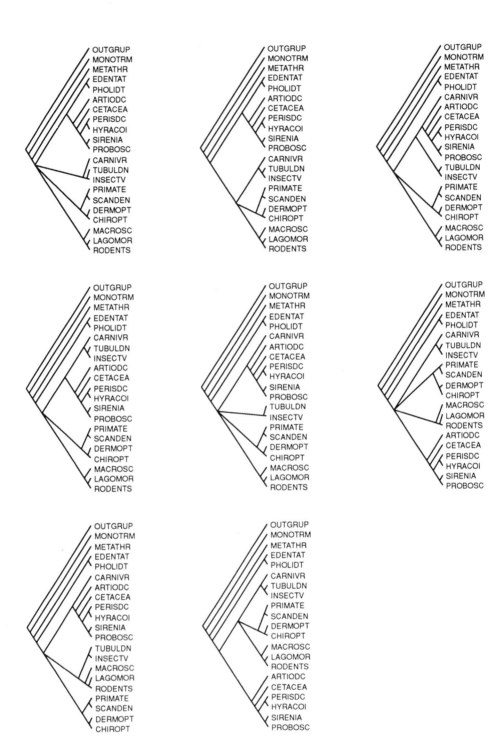

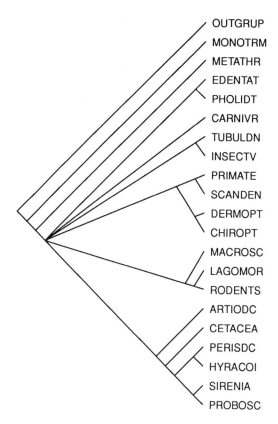

OUTGRUP
MONOTRM
METATHR
EDENTAT
PHOLIDT
CARNIVR
TUBULDN
INSECTV
PRIMATE
SCANDEN
DERMOPT
CHIROPT
MACROSC
LAGOMOR
RODENTS
ARTIODC
CETACEA
PERISDC
HYRACOI
SIRENIA
PROBOSC

Fig. 2.8. Strict consensus tree for trees displayed in figure 2.7. Acronyms as in figure 2.7.

FACING PAGE:

Fig. 2.7. Eight shortest trees for 20 orders of mammals using Recent data only. Trees are derived through use of PAUP version 3.0 (see Swofford 1985, for reference to earlier version) and HENNIG 86 (Farris 1988). Consistency (CI) and retention (RI) indices are 0.765 and 0.830, respectively. For data, see tables 2.1 and 2.2 and Novacek (1989). Acronyms are MONOTRM, Monotremata; METATHR, Metatheria; EDENTAT, Edentata; PHOLIDT, Pholidota; CARNIVR, Carnivora; TUBULDN, Tubulidentata; INSECTV, Insectivora; PRIMATE, Primates; SCANDEN, Scandentia (tree shrews); DERMOPT, Dermoptera (flying lemurs); CHIROPT, Chiroptera (bats); MACROSC, Macroscelidea; LAGOMOR, Lagomorpha; RODENTS, Rodentia; ARTIODC, Artiodactyla; CETACEA, Cetacea (whales); PERISDC, Perissodactyla; HYRACOI, Hyracoidea; SIRENIA, Sirenia (dugongs, manatees, sea cows); PROBOSC, Proboscidea.

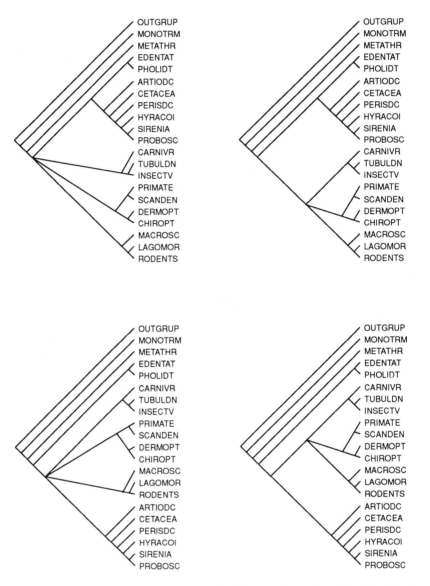

Fig. 2.9. Four most parsimonious trees produced from analysis of combined fossil and Recent data (tables 2.1 and 2.2) for 20 orders of extant mammals. Derived from use of both PAUP 3.0 and HENNIG 86. Consistency (CI) and retention (RI) indices are 0.768 and 0.842, respectively. Acronyms as in figure 2.7.

2.7. Figure 2.9 shows the four shortest trees derived from the combined fossil and Recent data set (tables 2.1 and 2.2), from which a consensus tree can be constructed (figure 2.10). In this tree Carnivora is grouped with Tubuliden-tata and Insectivora, whereas in figure 2.8 Carnivora is simply one of several branches in the major polytomous sector of the cladogram. A more interesting alteration that comes with the fossil ingroup data is the new position for the order Hyracoidea. In figure 2.10, the Hyracoidea has shifted from a sister group relationship with Perissodactyla (see fig. 2.7) to a closer relationship with Proboscidea and Sirenia. The latter two constitute, along with several fossil clades, the superordinal grouping Tethytheria (McKenna 1975). The group comprising both Tethytheria and hyracoids is commonly referred to as Paenungulata (Novacek et al. 1988) as modified from Simpson (1945).

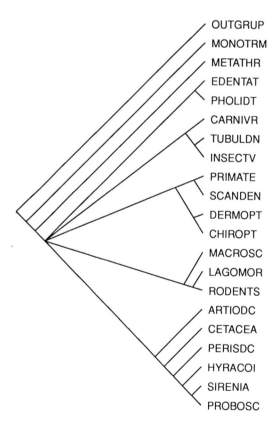

Fig. 2.10. Strict consensus tree for trees shown in figure 2.9. Acronyms as in figure 2.7.

The placement of the Hyracoidea is of some interest in higher mammalian phylogenetics. Alternatives shown in figures 2.8 and 2.10 mirror the debate between those advocating a tethythere-hyracoid grouping (Novacek and Wyss 1986; Shoshani 1986; Miyamoto and Goodman 1986; Novacek et al. 1988) and those supporting a perissodactyl–hyracoid grouping (Fisher 1986; Prothero et al. 1988). Critical support for the former argument is based on an accounting of early fossil perissodactyls, such as the Eocene *Hyracotherium*. This can be demonstrated by a revised analysis (tables 2.3 and 2.4; fig. 2.11) that (1) provides a more explicit treatment of the evidence for and against the Paenungulata; (2) identifies fossil perissodactyls (e.g., *Hyracotherium*) and fossil tethytheres as separate terminal taxa; and (3) accounts for unique traits shared by *Hyracotherium* and extant perissodactyls.

In the paenungulate problem, fossils show a particular combination of conditions that parallel those predisposing synapsids to affect the outcome of the amniote case. First, the fossil horses share autapomorphies with extant perissodactyls, like the saddle-shaped trochlea of the astragalus (Radinsky 1966). Second, the fossils are more primitive for certain traits shared by hyracoids and living perissodactyls. For example, modifications of distal phalanges common to the latter (Fisher 1986) are not found in *Hyracotherium*. At the same time, a close link between more recent perissodactyls and *Hyracotherium* forces a convergent pattern in distal phalangeal and other features that might otherwise appear as synapomorphies for living perissodactyls and hyracoids (fig. 2.11). Third, there is enough evidence for the grouping between hyracoids and tethytheres so that the addition of the fossils favors that grouping to the exclusion of perissodactyls (fig. 2.11). In other words, there is enough evidence for the monophyly of the hyracoid-tethy-

Table 2.3. Data Matrix for Hyracoid Relationships

	Characters 1111111112222222223 1234567890123456789012345678 90
Hyracotherium	0??00????00??01000000100000000
Perissodactyla	000000001111111000000100000000
Hyracoidea	111111111111111000000000000000
Sirenia	111101110000000111111000000000
Desmostylia	???10?????0??0011111?011000000
Proboscidea	111111110000000111111011111111
Moeritherium	???11????????0?11111?011111111

NOTE: For character explanations, see table 2.4. 0 = primitive condition; 1 = derived condition; ? = missing information due to preservational loss or marked transformation.

Table 2.4. Selected Morphological and Molecular Characters Relevant to the Problem of Interrelationships Among Hyracoid, Perissodactyls, and Tethytheres

---

1. Taxeopody (serial arrangement) of the carpal elements
2. Bifurcation of the musculus styloglossus
3. Zonary placenta
4. Amastoidy
5. Posterior extension of the jugal to the glenoid region
6. Replacement (position 110) α-hemoglobin
7. Replacement (position 44) β-hemogloblin
8. Replacement (positions 70, 74, 142) α-crystallin A
9. Musculus sternoscapularis present and inserts on the superior edge of the scapula
10. Acromiom process absent
11. Terminal phalanges broadened proximally and tapered distally, with median notch in the "nail bed"
12. Expanded eustachian sac
13. Extracranial course of the internal carotid artery
14. Development of the tuber maxillare in the molar region
15. Tridactyl pes
16. Forward displacement of the cheek teeth
17. Retracted nasals
18. Bilophodont crown patterns on cheek teeth
19. Infraorbital canal greatly reduced in length
20. Dorsally expanded process of the squamosal
21. Cochlear canaliculus absent
22. Saddle-shaped astragalonavicular articulation on astragalus
23. External auditory meatus high, nearly enclosed ventrally by mutual contact of the squamosal, posttympanic, and postglenoid processes
24. m3 hypoconulid complex, m3 with small entoconid II
25. Anterior jugal reduced
26. Absence (inferred loss) of orbitotemporal extension of the palatine
27. Absence (inferred loss) of condyloid foramen
28. I2, i2 enlarged
29. C,c reduced or lost
30. (D?)P1,(d?)p1 lost

NOTE: For character matrix, see table 2.3. For discussions of characters see Novacek and Wyss (1986), Fischer (1986), and Novacek et al. (1989).

there clade that the perissodactyls (including *Hyracotherium*) are not simply forced inside the clade.

The current status of the hyracoid debate is unsettled, with different perspectives on what characters are relevant and how they are described and evaluated (Prothero et al. 1988; Novacek 1989). The character set in tables 2.3 and 2.4 is likely to be revised, redefined, and expanded. The essential point here is that, under a given assumption about distributions of characters,

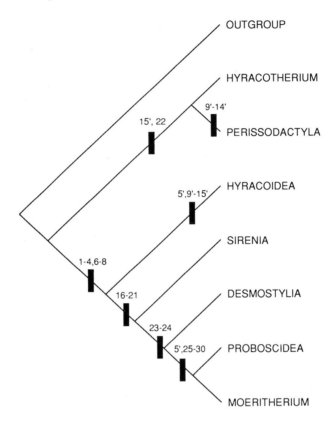

Fig. 2.11. Most parsimonious tree for revised analysis of hyracoid relationships. Fossil taxa *Hyracotherium, Moeritherium,* and Desmostylia are included. Tree is 38 steps in length with CI of 0.789 and RI of 0.826. Convergence or homoplasy in characters is indicated by apostrophes. Character state changes are optimized according to DEL-TRANS in PAUP 3.0. (For data see tables 2.3 and 2.4.)

fossils play a critical role in establishing the topology of relationships among these mammals.

Added taxa may, of course, affect phylogeny without drastically altering topology. Schemes for character transformation may be modified if added taxa destroy prior assumptions about the distribution of traits. Thus Donoghue et al. (1989) remarked that saccate pollen grains appear to be an autapomorphy of conifers when only extant seed plants are considered. When fossils are combined with the Recent data, saccate pollen is optimized as homologous for a large platyspermic clade that includes all extant taxa. The condition is retained in conifers but lost in ginkgos, cycads, and anthophytes. Likewise, the addition of the fossil *Hyracotherium* suggests that modification of terminal phalanges occurred independently in perissodactyls and hyracoids, even in the event that evidence might favor a close affinity between the latter to the exclusion of tethytheres (see figs. 2.7 and 2.8). Such added taxa introduce homoplasy where none may have been assumed. If one accepts the notion that age is directly related to retention of primitive traits, then addition of fossils is even more likely to disrupt patterns of homology derived strictly from extant forms. As shown by the amniote case (Gauthier et al. 1988), fossil taxa may even comprise a pectinate branching pattern that very clearly specifies a sequence of character change.

The Problem of Missing Data

Incomplete information is characteristic of poorly preserved fossils, but the problem is not confined to this category of data. Some molecular studies have applied "tandem alignments" for several proteins when not all proteins were known for all taxa considered. In the Miyamoto and Goodman (1986) study of several proteins in eutherian mammals, more than 30% of the characters in the data matrix could not be scored. Extant taxa may also elude scoring of certain characters because they show marked transformation. Gauthier et al. (1988) noted, for example, that the origin of the incus and malleus in mammals obscures any information on states representing the anlage to the ossicles—the lower jaw (postdentary) elements—in their synapsid relatives. Mammals must be scored as uninformative for these postdentary characters.

A feature of computer programs for cladistics, such as PAUP or HENNIG 86, is that they accommodate missing data. Missing data will be scored to conform with the character distributions favoring the most parsimonious solution. It must be remembered, nonetheless, that the algorithms employed by these programs merely accommodate missing data under a parsimony framework; they do not "solve the problem" of incomplete or uninformative data. Ambiguities, such as those noted earlier for fossil sister taxa (see fig. 2.4), are not necessarily erased when such algorithms are applied. More crucial, the assignment of some taxa based on incomplete data may simply be mistakes or artifacts that fail to represent the affinities of the taxon based on more complete information.

Another problem with these algorithms is that their accounting for missing entries is obviously not independent of the overall resolution of the tree. If the signal for a resolved tree based on known characters is weak, then addition of grossly incomplete fossils will hardly improve matters. Figure 2.12 shows what happens when seven fossil clades are added to the eutherian

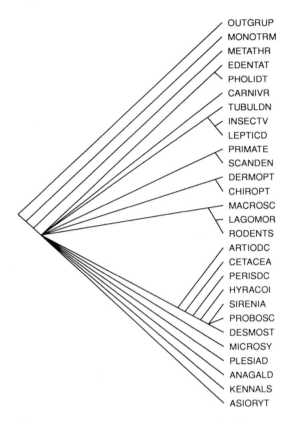

OUTGRUP
MONOTRM
METATHR
EDENTAT
PHOLIDT
CARNIVR
TUBULDN
INSECTV
LEPTICD
PRIMATE
SCANDEN
DERMOPT
CHIROPT
MACROSC
LAGOMOR
RODENTS
ARTIODC
CETACEA
PERISDC
HYRACOI
SIRENIA
PROBOSC
DESMOST
MICROSY
PLESIAD
ANAGALD
KENNALS
ASIORYT

Fig. 2.12. Analysis of mammal taxa incorporating seven fossil taxa as separate clades (see tables 2.1 and 2.2). Strict consensus solution of 6800 trees each of length 118 steps (CI = 0.75). Derived from use of PAUP 3.0. Acronyms other than those listed in figure 2.7 are DESMOST, Desmostylia; LEPTICD, Leptictida; MICROSY, Microsyopidae; PLESIAD, Plesiadapidae; ANAGALD, Anagalidae; KENNALS, Kennalestes, ASIORYTS, Asioryctes.

mammal data set. Without these fossils, the base of Eutheria is a poorly resolved polytomy resulting from conflicting most parsimonious solutions (see figs. 2.9 and 2.10). The fossils do affect the topology, primarily by fragmenting some of the clades preserved in figures 2.8 and 2.10. This change is, however, hardly salubrious, as the fossils merely serve to proliferate the number of most parsimonious trees. Figure 2.12 is thus a consensus of over 6800 shortest trees, which may not represent the limit of possible solutions (see comments in Novacek 1989). Similar complications result from the incorporation of poorly preserved Mesozoic fossil mammals in analysis of higher mammalian relationships (Rowe 1988; Greenwald 1989).

The preceding problem invokes a formula for exclusion of fossils or other taxa that show very poor information content. Rowe (1988), in fact, concluded that mammal fossils with less than 88% of the characters under study should be excluded from the analysis of higher mammal relationships. Although this prescription has some logic, it is often difficult to predict the effectiveness of taxa based on their amount of character information. In the amniote case, synapsids exhibited a wide range of effective preservation (fig. 2.13), but each of these taxa could serve equally well in altering the outcome of the analysis (Gauthier et al. 1988). In the hyracoid problem noted earlier, the fossil perissodactyl *Hyracotherium* is poorly represented by characters relative to both extant taxa and fossil desmostylians (fig. 2.14), yet this taxon bears critically on the analysis. The kinds of characters preserved, not just the degree of character representation, account for the potential influence of an added taxon. If among the few characters preserved are the combination of primitive and derived states that force relationships in a particular direction, then the included taxa—even when poorly represented—will play a significant role in the outcome. Of course, the possibility that these incompletely preserved taxa force the wrong outcome, which would be apparent had the taxa been better represented, cannot be eliminated.

Some of the fossil synapsid taxa in the amniote case are as impoverished as eutherian fossil taxa (compare figs. 2.13 and 2.15) considered in Novacek (1989). Why then were the latter fossils less critical? The answer seems more

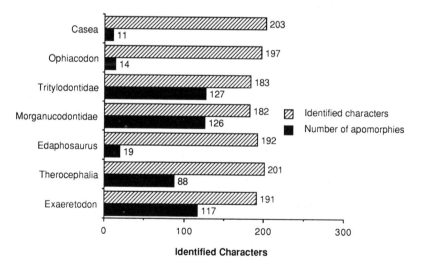

Fig. 2.13. Relative completeness of information represented by several synapsid taxa in the amniote case study. (Data from Gauthier et al. 1988).

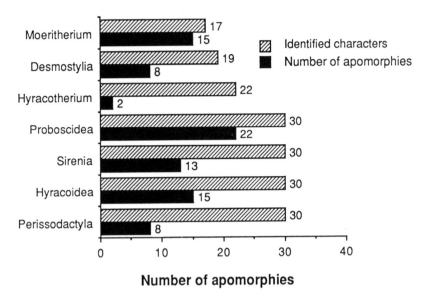

Fig. 2.14. Relative completeness of selected taxa bearing on the question of hyracoid relationships (see figure 2.11). (Data from tables 2.3 and 2.4.)

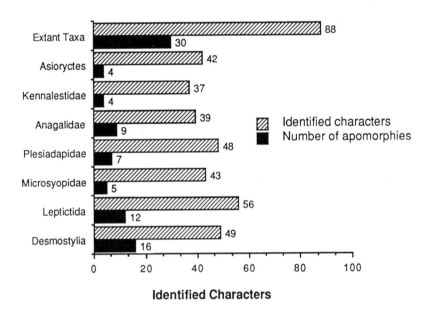

Fig. 2.15. Relative completeness of fossil taxa used in analysis of eutherian mammals (see figure 2.12). (Data from Novacek 1989; see tables 2.1 and 2.2.)

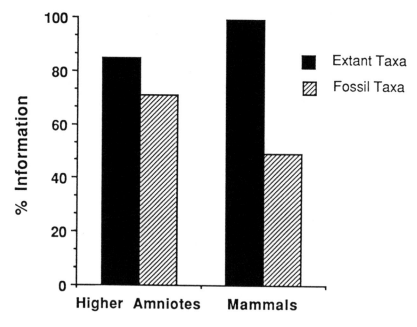

Fig. 2.16. Comparisons of completeness for fossil and extant taxa in the amniote case (Gauthier et al. 1988) and the eutherian mammal case (Novacek 1989). See text.

related to the information content of the extant taxa than the fossils. The amniote case is striking in that the extant taxa, especially mammals, have a large proportion of uninformative conditions. This ambiguity is due to marked transformation of numerous characters that apply to more basal nodes within the tree. The ratio of informativeness to uninformativeness (fig. 2.16) in the amniote case shows that ambiguity due to marked transformation is a salient feature of these data. Here fossil amniote clades compare favorably with extant clades in the percentage of informative characters. One would thus expect that fossils preserving critical combinations of traits exert a strong influence on the outcome of the analysis. In the eutherian mammal study, ambiguities that arise through transformation in extant taxa are not so evident. Only 1% of the data can be scored as uninformative on this basis (fig. 2.16). A large component of the character distribution is hence retrievable directly from study of extant taxa. Fossils, such as *Hyracotherium*, promote crucial but minor changes in this overall pattern. One might infer, then, that the evolution of extant eutherian orders subsequent to their split from other lineages is not so marked as to obscure data relevant to critical branching events.

Fossils, Relative Age, and Phylogeny

As noted earlier, fossil taxa have been highlighted as a special category of phylogenetic evidence on the assumption that antiquity is closely tied to primitiveness. How secure is the notion that older fossils tend to retain more primitive conditions than do their younger (sister group) relatives? An answer to this question in the form of a lawlike statement is not possible. Many ancient fossils are highly specialized forms and are sometimes even more transformed than either their less ancient or their living relatives. The fossil record can be very patchy, and patterns of character transformation may fail to emerge. Nonetheless, it is claimed that the study of fossils provides direct insight into the polarity of character change. In answer to the charge that ancientness and primitiveness are not necessarily correlated, Simpson (1975: 14) remarked, "That is true but they *usually* are correlated, and for any group with even a fair fossil record there is seldom any doubt that characters usual or shared by older members are almost always more primitive than those of later members."

Lately these arguments have been subjected to analytic tests. Gauthier et al. (1988: fig. 5) recorded a very close match between the age of an amniote taxon (documented by its first appearance in the fossil record) and its cladistic rank (i.e., the number of nodes a given taxon is removed from the most recent common ancestor of all extant amniotes). The correlation was significant at $P > 0.05$, whether the cladogram of Gardiner (1982) or Gauthier et al. (1988) was used to order cladistic rank. Obviously, the degree of resolution of the cladogram, regardless of the particular topology favored, will affect correlations between age and ranking. In pectinate cladograms, rank ordering of cladistic events shows a linear increase; in "bushy" cladograms, it shows a stepwise increase. The unresolved polytomy near the base of the eutherian tree (e.g., fig. 2.10) thus strongly contracts cladistic rankings. For those components of figure 2.10 that are arrayed along the main axis of the tree, the correlation between age and cladistic rank is diminished in part by the lack of resolution in cladistic patterns (fig. 2.17). For example, insectivores, primates, and rodents are each the oldest representatives of superordinal clades that diverge from an unresolved polytomy. Points representing these groups in figure 2.17 are accordingly aligned parallel to the horizontal ("age rank") axis of the graph.

Another factor that decreases the fit between the age and cladistic rank is the lack of resolution in the record of first occurrence of the groups in question. Thus, monotremes, insectivorans, and primates are represented by points aligned parallel to the vertical axis of the graph (fig. 2.17), and several other sets of points are aligned in a similar fashion. This is not surprising;

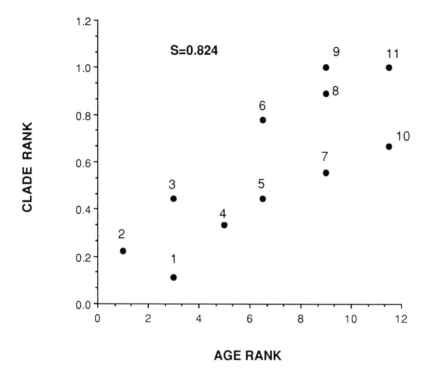

Fig. 2.17. Correlation between age rank and cladistic rank for mammal groups resolved in figure 2.10. Age ranks are ordered and ties are averaged from raw age data. Clade ranks are rescaled between 0 and 1. Because of age-scaling problems, not all taxa shown in figure 2.10 are ranked; only taxa forming pectinate components or in polytomies along the main axis of the cladogram are ranked. S = Spearman's rank correlation coefficient. Numbers refer to the following taxa: 1, Monotremata; 2, Metatheria (marsupials); 3, Insectivora; 4, Edentata; 5, Rodentia; 6, Perissodactyla; 7, Artiodactyla; 8, Hyracoidea; 9, Proboscidea; 10, Cetacea; 11, Sirenia.

mammal group durations based on first appearances in the fossil record can be extended backward in time when lineage durations based on age of the nearest sister taxa are considered (see Norell, chapter 3 of this volume). This operation shifts the age ranking of certain taxa to a better alignment with their cladistic ranking. In the mammal case, the raw datum for group age—namely, the first appearance of the mammalian order in the fossil record—is not necessarily a reliable predictor of the degree of "primitiveness" of the taxon relative to other orders. This seems due to both poor resolution at the base of the eutherian tree and the existence of several orders whose known fossil records grossly underrepresent the age of those orders. On the other hand, fossils assigned within these orders might better reveal close correlations between age and primitiveness. For example, the earliest perissodactyl, *Hyracotherium*, is the taxon least removed from the basal node of the order (Radinsky 1966). Similar correlations might be found in orders with either or both relatively rich fossil records and well-represented "plesiomorphous" taxa (e.g., Carnivora, Rodentia, Primates, Insectivora, Artiodactyla, Proboscidea, and Chiroptera), although data have not yet been compiled. Indeed, a general comparative study of known age (first appearance) data with the recent spate of cladistic analyses would be instructive. Although there is reason to claim that fossil age data in some general sense conform to the primitive-derived spectrum for characters, this assertion warrants scrutiny through comparisons of range data and cladistic rankings in diverse taxa (see also chapter 3, this volume).

Criteria for Evaluating Trees

Cladistic analysis is aimed at deriving a tree that best fits observed character distributions. Under the principle of parsimony, the preferred tree enlists the fewest number of steps to explain the character data. Derivation of the most parsimonious tree may not always be deemed a satisfactory solution to a question involving relationships. If, as noted earlier, the shortest trees show polytomous branching, resolving power of the data is judged deficient. Moreover, a tree may not represent all its monophyletic groupings with comparable effectiveness. A group that collapses in a tree that is one or two steps longer than the most parsimonious cladogram may be respected less than a group showing a higher degree of stability when more homoplasy is allowed. Histograms that show distributions of tree lengths for all possible trees (a subroutine in PAUP 3.0) provide a basis for comparative evaluation of trees and the groups they identify. In the hyracoid problem, a histogram for all trees of different lengths (fig. 2.18) derived from the data set in table 2.3 shows that collapse of Paenungulata occurs only three steps away from the single most parsimonious tree (fig. 2.11). In contrast, Tethytheria collapses in trees at least seven steps longer than the most parsimonious tree (fig. 2.18). Statistical tests for trees showing taxa at differing positions have been suggested (Templeton 1983; Hillis and Dixon 1989). Eschewing such tests, we are left with the impression that Tethytheria—under the criterion of stability—is a more cohesive grouping than Paenungulata. Tethytheres are deemed more likely to survive tests involving additions of characters and or taxa than are paenungulates (sensu Novacek et al. 1988).

Comparisons of trees of different lengths have found a particular use in

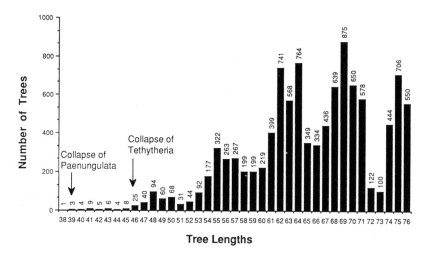

Fig. 2.18. Histogram for trees of all lengths derived from data in table 2.3. Arrows indicate collapse of groups preserved in the most parsimonious tree (figure 2.11).

studies involving large data sets, such as those that characterize molecular evidence. Miyamoto (1989) made exhaustive comparisons of 10,395 trees derived from mitochondrial DNA sequences for both rRNA mt genes in artiodactyls (fig. 2.19). The most parsimonious tree identified a grouping of *Bison bison* and *Bos grunniens* that remains intact until one encounters trees at least 20 steps longer, indicating an association between *Bos grunniens* and *Bos taurus*. The *Bison–Bos* clade thus shows a level of stability with increasing homoplasy that notably surpasses the paenungulate grouping shown in figure 2.11.

Applying such a criterion to the evaluation of different case studies, however, can be problematic. An obvious concern is the disparity in the size of different data sets. One might expect that groups be retained for 20 or more steps, where the number of characters evaluated, as in Miyamoto's (1989) exemplary study, are hundreds of informative base positions along the DNA molecule.

Effects of character sample size are irrelevant to another problem in drawing inferences from the relative stability of different groupings. Consider a data set comprising five characters for three taxa (fig. 2.20). Four characters are shared by group BC, one unique character is present in C, and no characters are shared by any other combination of groups. For three possible outcomes, group BC shows appreciable stability. It takes four steps to collapse BC and yield the alternatives. A taxon (X) is added that shares two of the four

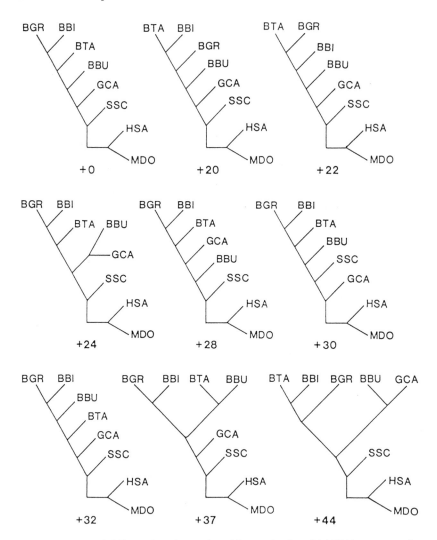

Fig. 2.19. Trees of different lengths produced from mitochondrial DNA sequences for selected artiodactyl species. (From Miyamoto and Boyle 1989.)

characters of BC and shares character 5 with C (fig. 2.21). Here the shortest topology for ABC matches that shown in figure 2.20. The favored grouping of BC is, however, more unstable; it takes only three steps to collapse BC when X is included. Clearly, this is because X introduces homoplasy in the form of a character shared exclusively with C. At the same time, X provides some additional information not retrievable from the character matrix limited to A, B, and C, namely, that characters 3 and 4 are expressed at a more general level than are characters 1 and 2. In addition, the shortest tree that includes X suggests that character 5 either is independently acquired in C and X or is a general condition for B, C, X (see fig. 2.21). By contrast,

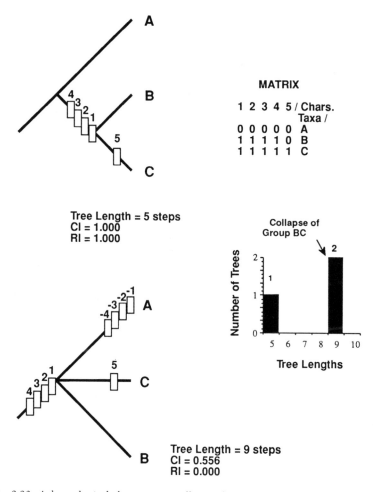

Fig. 2.20. A hypothetical data matrix, "all-tree" histogram, and alternative trees for taxa A, B, and C. Histogram shows number of dichotomous trees only. The two dichotomous trees of nine steps (see histogram) each have a zero-length branch, which, when collapsed, produces the single topology shown below. CI = consistency index; RI = retention index.

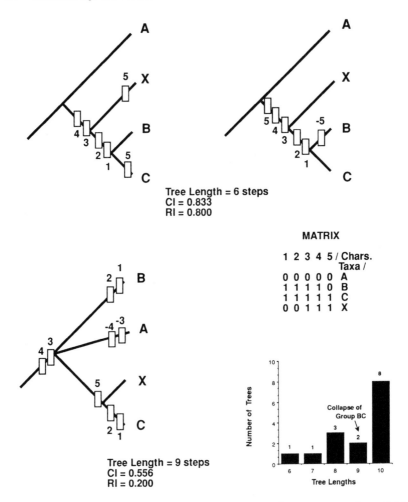

Fig. 2.21. Same data as in figure 2.20 with the addition of taxon X. As in figure 2.20, Histogram shows dichotomous trees only. The shortest topology is shown above with two alternative optimizations for character 5. The two trees of nine steps (see histogram) each have a zero length branch, which, when collapsed, yields the single topology shown below.

character 5 is represented simply as an autapomorphy in C in the shortest tree for A, B, C (see fig. 2.20). The presence of character 5 in taxon X introduces homoplasy on the shortest tree C(AB), but this homoplasy is critical to a revised understanding of the level of generality at which character 5 is expressed.

These hypothetical examples suggest a general observation: cladograms with fossils or extant taxa functioning as "transitional" taxa, such as taxon X in figure 2.21, will tend to produce groupings of lower stability (with increasing homoplasy) than cladograms in which such taxa are rare or lacking. In the first category of cladograms, nodes that might be supported by several characters must share these characters with several other nodes. In a paleontological context, cladograms that include fossil transitional taxa can be more unstable than those for the same extant taxa that lack the fossils. But this seemingly unfavorable quality collides with an inherent attribute: the additional taxa not only enrich the taxonomic diversity of a cladogram, but also provide more detailed information on the pattern of character transformation. Thus a maximally resolved tree under one criterion—namely, a tree where every evolutionary step is represented by a taxon—might be judged, in some instances, as an inferior result under a criterion that emphasizes stability in the face of increasing homoplasy. Comparisons of trees of varying lengths may provide some means for gauging the relative strengths of different phylogenetic hypotheses, but such comparisons may also minimize the importance of transitional taxa represented by fossils.

The Importance of Fossils

Any taxon added to a problem set has the potential to shift topologies or alter schemes of character transformation. This potential will be greater if the cladogram prior to the addition of taxa fulfills certain conditions. As argued here and elsewhere, such vulnerable cladograms will tend to show an accumulation of traits at particular nodes, a balance of evidence that only slightly favors a critical hypothesis over an alternative one, and a branching pattern that shows a high degree of resolution in the form of pectinate trees. Moreover, an added taxon will most likely modify topology if it introduces character conflict, primarily through a combination of derived and primitive traits. The derived traits convert former autapomorphies into synapomorphies by supporting a special relationship between the added taxon and one or a few taxa in the original cladogram. The primitive traits isolate the added taxon, assuming these traits are modified in the other taxa of the original tree. The combination of traits in the added taxon serve to detach a related group from its former position in the original cladogram. If, for instance, the weight of primitive traits in the added taxon is high, the affected group will be shifted to a more basal position.

Although extant taxa as well as fossils may alter cladograms that fulfill the preceding conditions, fossils will have a particular tendency to affect topology and character change if a basic correlation between age and primitiveness is upheld. There are, accordingly, examples where fossils indicate the geometry of splitting events that are not retrievable from extant taxa alone. As case studies accumulate, the prevalance of such examples will be better understood. The potential of fossils to provide transitional taxa that more explicitly describe a sequence of character change is also recognized, even though cladograms that accommodate such transitional taxa may have highly unstable sectors.

This review is developed along a particular theme. It is hoped that we are beyond any debate as to whether fossils are of primary importance in questions involving the relationships of extant taxa. The answer is, of course, that they are always important, and the conditions under which they seem especially critical have been described. It seems now more productive to look to those analytical problems (such as variably represented data, polymorphism, and evaluation criteria) that impede our progress to the overall goal of those who study fossils and extant taxa: phylogenetic reconstruction.

ACKNOWLEDGMENTS

I thank Gavin Naylor, Steve Farris, Tim Rowe, Nancy Simmons, Sheri McGehee, Greg Edgecombe, Andy Wyss, Ward Wheeler, Michael Donoghue, Kevin Nixon, Jacques Gauthier, Quentin Wheeler, Martin Fischer, Rodney Honeycutt, Ed Wiley, Dave Canatella, Jim Carpenter, Paul Vrana, Rick Harrison, and students of the Cornell University Evolutionary Biology Program for discussions. For critical readings, I thank Mark Norell, Greg Edgecombe, Quentin Wheeler, and Malcolm C. McKenna.

REFERENCES

Ax, P. 1985. Stem species and the stem lineage concept. *Cladistics* 1:279–287.
Darwin, C. 1859. *The Origin of Species*. New York: Mentor (1958).
Donoghue, M., J. Doyle, J. Gauthier, A. Kluge, and T. Rowe. 1989. The importance of fossils in phylogeny reconstruction. *Ann. Rev. Ecolo. Syst.* 20:431–460.
Doyle, J. and M. Donoghue. 1986. Seed plant phylogeny and the origin of angiosperms: An experimental cladistic approach. *Bot. Rev.* 52:321–431.
Doyle, J. and M. Donoghue, 1987. The importance of fossils in elucidating seed plant phylogeny and macroevolution. *Rev. Paleobot. Palynol.* 50:63–95.
Farris, J. S. 1988. *HENNIG 86*, Version 1.5. Distributed by the author. 41 Admiral St., Port Jefferson Station, New York.
Fischer, M. 1986. Die Stellung der Schliefer (Hyracoidea) im phylogenetischen System der Eutheria. *Cour. Forsch. Inst. Senckenberg* 84:1–132.

Gardiner, B. 1982. Tetrapod classification. *Zool. J. Linn. Soc.* 74:207–232.

Gauthier, J., A. Kluge, and T. Rowe. 1988. Amniote phylogeny and the importance of fossils. *Cladistics* 4:105–209.

Gingerich, P. and M. Schoeninger. 1977. The fossil record and primate phylogeny. *Jr. Hum. Evol.* 6:482–503.

Goodman, M. 1989. Emerging alliance of phylogenetic systematics and molecular biology: A new age of exploration. In B. Fernholm, K. Bremer, and H. Jornvall, eds., *The Hierarchy of Life,* pp. 43–61. New York: Elsevier.

Greene, J. 1961. *The Death of Adam.* New York: Mentor.

Greenwald, N. 1989. Effects of missing data and homoplasy on estimates of multituberculate phylogeny. *J. Vertebr. Paleontol.* (abstracts), 9:24A.

Hennig, W. 1966. *Phylogenetic Systematics.* Urbana: University of Illinois Press.

Hennig, W. 1981. *Insect Phylogeny.* New York: John Wiley.

Hillis, D. and M. Dixon. 1989. Vertebrate phylogeny: Evidence from 28S ribosomal DNA sequences. In B. Fernholm, K. Bremer, and H. Jornvall, eds., *The Hierarchy of Life,* pp. 355–367. New York: Elsevier.

Jefferies, R. 1979. The origin of chordates—a methodological essay. In M. R. House, ed., The origin of major invertebrate groups. *Syst. Assoc. Special.* 12:443–447. London: Academic Press.

Løvtrup, S. 1985. On the classification of the taxon Tetrapoda. *Syst. Zool.* 34:463–470.

McKenna, M. C. 1975. Toward a phylogenetic classification of the Mammalia. In W. P. Luckett and F. S. Szalay, eds., *Phylogeny of Primates,* pp. 21–46. New York: Plenum Press.

McKenna, M. C. 1987. Molecular and morphological analysis of high-level mammalian interrelationships. In C. Patterson, ed., *Molecules and Morphology in Evolution: Conflict or Compromise?* pp. 55–93. Cambridge: Cambridge University Press.

Miyamoto, M. and S. Boyle. 1989. The potential importance of mitochondrial DNA sequence data to eutherian mammal phylogeny. In B. Fernholm, K. Bremer, and H. Jornvall, eds., *The Hierarchy of Life,* pp. 437–450. New York: Elsevier.

Miyamoto, M. and M. Goodman. 1986. Biomolecular systematics of eutherian mammals: Phylogenetic patterns and classification. *Syst. Zool.* 35:230–240.

Nelson, G. 1978. Ontogeny, phylogeny, paleontology and the biogenetic law. *Syst. Zool.* 27:324–345.

Novacek, M. 1989. Higher mammal phylogeny: The morphological-molecular synthesis. In B. Fernholm, K. Bremer, H. Jornvall, eds., *The Hierarchy of Life,* pp. 421–435. New York: Elsevier.

Novacek, M. and M. Norell. 1982. Fossils, phylogeny, and taxonomic rates of evolution. *Syst. Zool.* 31:366–375.

Novacek, M. and A. Wyss. 1986. Higher-level relationships of the Recent eutherian orders: Morphological evidence. *Cladistics* 2:257–387.

Novacek, M., A. Wyss, and M. McKenna. 1988. The major groups of eutherian mammals. In M. Benton, ed., *The Phylogeny and Classification of the Tetrapods, Vol. 2: Mammals,* pp. 31–71. Oxford: Clarendon Press.

Patterson, C. 1981. Significance of fossils in determining evolutionary relationships. *Ann. Rev. Ecol. Syst.* 12:195–223.

Prothero, D., E. Manning, and M. Fischer. 1988. The phylogeny of ungulates. In M. Benton, ed., *The Phylogeny and Classification of the Tetrapods, Vol. 2: Mammals,* pp. 201–234. Oxford: Clarendon Press.

Radinsky, L. 1966. The adaptive radiation of the phenacodontid condylarths and the origin of the Perissodactyla. *Evolution* 20:408–417.

Rowe, T. 1988. Definition, diagnosis, and origin of Mammalia. *J. Vertebr. Paleontol.* 8(3):241–264.

Sanderson, M. and M. Donoghue. 1989. Patterns of variation in levels of homoplasy. *Evolution* 43(8):1781–1795.

Shoshani, J. 1986. Mammalian phylogeny: Comparison of morphological and molecular results. *Mol. Biol. Evol.* 3:230–240.

Simpson, G. 1945. The principles of classification and a classification of mammals. *Bull. Amer. Mus. Nat. Hist.* 85:1–350.

Simpson, G. 1961. *Principles of Animal Taxonomy.* New York: Columbia University Press.

Simpson, G. 1975. Recent advances in methods of phylogenetic inference. In P. Luckett and E. Delson, eds., *Phylogency of the Primates,* pp. 3–19. New York: Plenum Press.

Swofford, D. 1985. *PAUP: Phylogenetic Analysis Using Parsimony.* Illinois Nat. Hist. Survey, 1985–1990.

Templeton, A. 1983. Phylogenetic inference from restriction endonuclease cleavage site maps with particular reference to the evolution of humans and the apes. *Evolution* 37:221–244.

3 : Taxic Origin and Temporal Diversity: The Effect of Phylogeny

Mark A. Norell

Abstract. Study of origin and diversification of organisms has been based on their first occurrence and duration in the fossil record. This research has progressed without a phylogenetic component. Phylogenetic systematics, provides a framework that enhances patterns seen in the fossil record by extending the history of taxa in reference to calibrated phylogenetic hypotheses. From these hypotheses lineage durations can be extracted. A lineage represents the minimum temporal duration of a taxon since it split from its sister group. Lineage durations can be integrated into studies of temporal origination and evolutionary rate and have the effect of increasing durations of taxa and modifying origination histories. No general rules predict the fashion in which phylogenetic correction of temporal ranges modifies origination or duration. Cases must be examined individually, since cladistic structure and distribution of fossils are influential. A second type of analysis uncovers ghost lineages in a phylogeny's internal structure. Ghost lineages and lineages show that actual (phylogenetically based) temporal diversities are counterintuitive when compared with the fossil record. This principle is illustrated in reference to the origination of eutherian mammal orders. A phylogenetic approach may be superior to gross accumulation of fossil first occurrences and durations because it interprets the fossil record relative to testable phylogenetic patterns.

Estimating first occurrence and temporal duration of taxa is the key element in determination of taxonomic evolutionary rates (Simpson 1944, 1953; Raup and Marshall 1980), origin and demise of faunas (Sepkoski 1988; Raup and Boyaijian 1988), and evolutionary modes and processes associated with taxonomic evolution (Vrba 1985; reviewed in Raup and Jablonski 1986). A critical weakness of such studies is their empirical base—a total reliance on first and last occurrence in the known fossil record for estimating time of origin and temporal duration of taxa.

Determination of timing of phyletic radiation and temporal diversity and calculation of taxonomic evolutionary rates have developed without a phylogenetic component (see remarks in Cracraft 1981, 1985; Eldredge and Novacek 1985) for determining temporal duration of taxa and their time of origin. I will show that a phylogenetic approach can alleviate some problems caused by literal interpretation of an inconsistent and inadequate fossil record.

The method of determining minimum time of origin of taxonomic groups is imprecise. The usual practice is to extend the known range of a taxon into the past with a dotted line. The length of the line is usually based on subjective criteria that estimate the length of time required by a hypothesized evolutionary process to form a certain morphotype. Such studies lack a formal procedure or testable basis for determining minimum or probable first temporal occurrence of a group—data from which taxonomic rates of evolution should be calculated and the framework within which patterns of phyletic origin should be phrased.

Phylogenetic systematics offers an alternative approach: one that takes the known fossil record and phylogeny of a group and uses oldest known fossils of each sister group to arrive at a minimum time of divergence for both groups. This approach poses the question, "Are there phyletic precursors that extend the histories of fossil groups?" I propose that taxonomic groups have invisible, missing segments of temporal history that significantly extend their temporal ranges. I also argue that additional taxa are sometimes buried in the structure of phylogenetic trees. These antecedents to fossil first occurrences and these additional taxa are the unrecognized predictions of phylogenetic trees that can be identified through a synthesis of the fossil record and phylogenetic systematics.

When temporal durations of taxa are significantly modified and additional "invisible taxa" are discovered, the data set customarily used to examine several contemporary issues in paleobiology and evolutionary biology is shown to be deficient. At best this deficiency will require slight adjustments; at worst a rethinking of historical patterns and their associated processes may be necessary.

Methods

Although it has been recognized that phyletic splitting may predate the acquisition of diagnostic characters of descendant groups, this realization has not been formalized or incorporated by paleontologists into the study of temporal origination. The most rigorous development of this issue is Hennig's (1965). He explicitly recognized that phyletic origin is the splitting of an evolutionary line. This splitting can predate the origin of a group's "typical" characters. I use *typical* as equivalent to *diagnostic* in the preferred sense of Rowe for characters that "are hypothesized to have been evolutionary novelties in *the* most *recent* common ancestor" (Rowe 1987:210). Hennig's ideas concerning the relationship between cladogenesis and character acquisition are indicated in figure 3.1. Hennig recognized three events in the origin of a phylogenetic group: T-1, the time of origin of a group; T-2, the time of acquisition of a group's typical characters; and T-3, the time of a group's diversification.

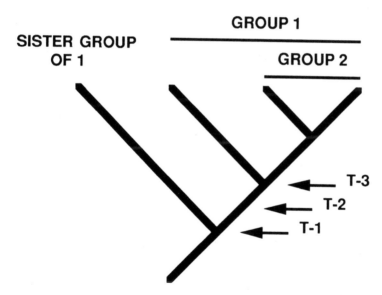

Fig. 3.1. The three time periods in the origin of a phylogenetic group recognized by Hennig (1965). T-1 is the origin of the group, T-2 is the time of acquisition of the group's typical characters, and T-3 is the time of diversification. (Redrawn from Hennig 1965.)

These events are difficult to determine empirically using Hennig's distinction. Hennig defined T-1 as the point at which an evolutionary line splits and its descendants begin to pursue independent evolutionary fates. However, until one of the lines achieves T-2, no evidence of splitting is present, because diagnostic characters have yet to appear. Until diagnostic characters appear in one sister lineage, the taxa are indistinguishable and appear as a single group in the fossil record, unchanged from the ancestral condition (fig. 3.2). If sister taxa are preserved in the fossil record as contemporaries, one apomorphic taxon demonstrates the presence of two evolutionary lines by default since the plesiomorphic taxon is excluded from ancestry.

Because characters that diagnose a group may originate at any time between the origin of a lineage and the first fossil occurrence of a taxon with diagnostic apomorphies, the age of a group is not necessarily synonymous with the time it phyletically split from its sister taxon. In describing the difference between T-1 and T-2 Hennig was in effect demonstrating the difference between taxonomic and morphologic evolution. These concepts, when confused and combined with an inconsistent fossil record, have hampered studies dependent on the documentation of the temporal history of cladogenesis.

To determine the minimum age of a taxon empirically requires that phyletic and morphologic evolution be separated and that the age of a taxon be defined as its entire temporal duration since it split from its sister taxon. To facilitate this discussion I refer to two conventions:

A *group* at any level is diagnosed only by apomorphy. A group's first temporal occurrence is documented only by the occurrence of apomorphy (i.e., diagnostic characters) in a fossil taxon.

A *lineage* is the entire temporal extent of a phyletic branch giving rise to a monophyletic group or a branch excluded from ancestry by temporal synchrony.

Groups are the traditional commodity of paleobiologists, and their first occurrences are read directly from their occurrence in the fossil record. Lineages represent total durations of discrete evolutionary lines and become groups with the development of diagnostic characters. The concept of lineages subsumes that of groups. The firsts occurrence of a lineage is recognized by either the oldest member of their subsumed group or the most ancient member of their immediate sister group. Minimum lineage ages always equal or exceed minimum group ages. This approach discovers parts of phylogenetic history that are not considered by traditional methods (the difference between lineage and group age), because they are segments of phyletic history invisible to the fossil record.

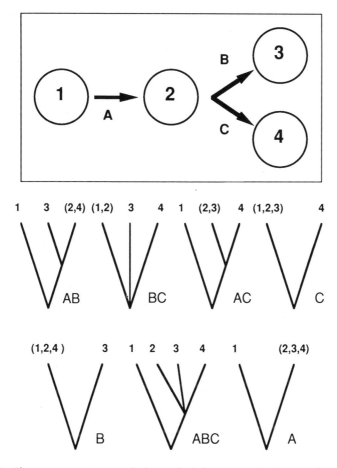

Fig. 3.2. Character acquisition and observed phylogeny in the history of a clade. A population (represented by circles) undergoes a history of anagenic change and cladogenesis. Character acquisition (Hennig's T-2, which allows populations to be differentiated) is indicated by capital letters. Phylogenetic hypotheses reflect all possible permutations of patterns of character acquisition in the clades' history. In, for example, phylogeny AB, characters are acquired during the times A (the transition from 1 to 2) and B (the transition from 2 to 3). Because no character change takes place during temporal segment C, 2 and 4 are indistinguishable as separate groups. Depending on pattern of character acquisition, recognition of groups and phylogenetic history are variable.

Throughout the test I refer to these conventions in the following fashions. *Dg* is the minimum group duration determined by a group's first and last appearance in the stratigraphic record. *Dl* is the minimum duration of a lineage. *Ag* is the minimum age of a group's first occurrence determined from its initial appearance in the fossil record, and *Al* is the minimum first occurrence of a lineage.

Fossil records and phylogenetic hypotheses can be combined into calibrated phylogenetic hypotheses from which Dg, Dl, Ag, and Al can be empirically determined for all included taxa (fig. 3.3). This procedure requires that (1) phylogenetic hypotheses be based on nested sets of derived (= apomorphic) characters, (2) all taxa be strictly monophyletic (in the sense of Donoghue 1985), and (3) specimens be referred to groups solely on the basis of diagnostic characters.

Operationally, this procedure uses a phylogenetic branching diagram and a fossil record to produce a calibrated phylogeny (see fig. 3.3), observing that sister lineages must have equal minimum temporal durations. Branch points are given minimum temporal ages equivalent to the oldest member of each pair of sister taxa.

The monophyly criterion is crucial. (I use *monophyly* in the restricted sense of Donoghue 1985). This definition suggests that taxa (including species) can only be allocated to bona fide monophyletic taxa on the basis of derived characters. Donoghue's conception of monophyly is important because when applied to species it explicitly requires them to possess unique apomorphies. If only monophyletic taxa are considered, the minimum temporal history of an entire lineage from its origin at a branch point can be determined. In some cases nonmonophyletic (i.e., the metaspecies of Donoghue 1985 and Gauthier 1984 as adapted by Norell 1989) species can be used in calibrated phylogenetic trees. For instance, in figure 3.3, if taxon D were nonmonophyletic, it would have a length of 4. The appearance of the diagnostic taxon C indicates that all members of the D lineage after 4 compose at least one separate lineage. These coeval D taxa are excluded from the ancestry of C; those that occur before the first appearance of C may be construed as putative C ancestors.

The differences between the evolutionary history of a group presented by a literal interpretation of the fossil record and that portrayed by calibrated phylogenetic hypotheses can be empirically compared between clades or among different phylogenetic topologies. This can be expressed through the index Z:

$$Z = 1 - \frac{\Sigma \ (Al - Ag)/\text{number of taxa}}{Ag \text{ of clade}}$$

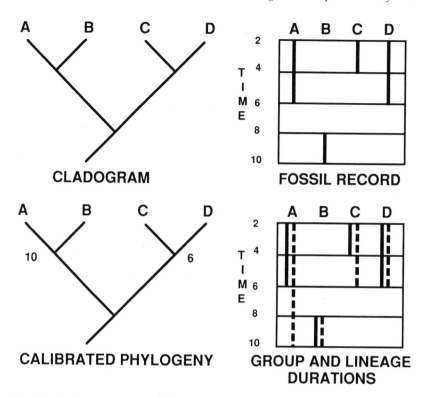

Fig. 3.3. Cladistic structure of monophyletic taxa can be combined with observed stratigraphic ranges to develop a calibrated phylogeny observing the convention that sister groups are equivalent in age. If relationships in the cladogram are preferred, group AB has a minimum origin at time 10, because group B first appears during this interval. The origin of group B must postdate the origin of lineage A because its diagnostic characters exclude it from the ancestry of A. From the calibrated phylogeny the durations of lineages (indicated by dashed lines) can be extracted. Lineage ranges and durations will always be longer than or equal to the durations and ranges of groups (solid bars).

where Ag of clade is determined by the oldest member of the clade. Z is a comparison between the total amount of time represented by lineages and the time represented by groups and is used here as an indicator of fit between fossil records and phylogenies determined from cladistic analyses. Z indicates the amount of discrepancy between fossil records and phylogenies and is useful in illustrating lineage and group inequality. The value of Z varies between 0 and 1. A value of 1 indicates isometry between ages of phyletic lines and ages of groups and a value of 0 corresponds to lack of a fossil record. For Z to approach 1 all pairs of sister taxa must appear nearly simultaneously in the stratigraphic record.

It is doubtful that Z will ever closely approach 1, because of sampling (Schindel 1980; Dingus and Sadler 1982), preservational vagaries, and taphonomic (i.e., preservational) processes and the fossil record's inaccuracy in portraying evolutionary events. Figure 3.2 demonstrates scenarios where phyletic splitting occurs but remains unrecognized because one of the lineages fails to evolve diagnostic apomorphies. This diagram depicts one manifestation of Hennig's deviation rule. This result is also predicted by several current theories of evolutionary change (Eldredge and Gould 1972; Brooks and Wiley 1986; Vrba personal communication) in which the formation of a group (the evolution of diagnostic characters) occurs instantaneously in geological time but asymmetrically in lineages after phylogenetically branching.

The equivalence of taxonomic duration and lineage duration determined from the age of sister groups is the first dimension of this study. In effect, this procedure phylogenetically corrects observed fossil first occurrences and durations.

These corrected durations (Al and Dl) are better data for examination of phylogenetic tempos and patterns than a literal reading of the fossil record, because they reflect a synthesis of phylogeny reconstruction and the stratigraphic record. The corrected data can be applied to two examples that represent common themes in contemporary paleobiology. In these, no definitive conclusions are expressed; instead the examples show how conclusions are modified by implementing a phylogenetic approach.

Example 1: Phylogenetic Hypotheses and Evolutionary Tempo

For more than 50 years paleobiologists have been preoccupied with the estimation of evolutionary rates. In their crudest form, these estimates are qualitative measures of relative tachytely and bradytely (respectively, fast and slow taxonomic evolutionary rates) among groups (Simpson 1944). However, methodologies to estimate more universal quantitative measures have also been proposed (e.g., Kurtén 1960; Stanley 1979; and Levinton and Farris 1987, among others). One commonly used method of estimating taxonomic evolutionary rates is construction of survivorship curves and extraction of a rate from a slope function (Van Valen 1973).

Here I examine the difference between estimates of taxonomic longevity derived from a traditional data set and estimates derived from the same procedure after its phylogenetic correction. Again, although other estimating procedures are commonly applied, they utilize the same data set. As pointed out by Novacek and Norell (1982), lengthening the temporal ranges of taxa influences results from these procedures in a similar fashion: an increase of the duration of evolutionary units will modify rates.

To examine the magnitude of a phylogenetic correction of duration on survivorship curves, an artificial data set was generated (fig. 3.4). Temporal ranges of 10 monophyletic taxa are shown by vertical bars. Four widely discrepant tree topologies (see fig. 3.4) were analyzed to detect Dls, following the procedure outlined in figure 3.3. For each phylogeny Dls were determined for individual taxa. Survivorship curves were constructed (fig. 3.5) from these Dls and Dgs (the traditional data for such studies). The large number (34,459,425) of possible dichotomous phylogenetic hypotheses for 10 taxa (Felsenstein 1978) disallows analysis of all but a small sample, yet the four selected trees probably represent extremes cases. The range of possible curves is approximately demonstrated by curves B and C in figure 3.5. The differences are due to variation in the cladistic structure of the calibrated phylogenies (see fig. 3.4). Curve A ($Z = 0.85$) is derived from a cladogram with a young terminal taxon (E) whose sister group is ancient (D). Curve C ($Z = 0.93$) is derived from a calibrated phylogeny that is approximately compatible with the fossil occurrences of the taxa in figure 3.4. Dls extracted from phylogeny B ($Z = 0.65$) were widely discrepant from the fossil record. Phylogeny D has a Z of 0.86 and indicates a survivorship pattern similar to that derived from phylogeny A.

The diversity of survivorship curves shows that disparate pictures of evolutionary tempo and taxonomic proliferation are predicted by different phylogenetic interpretations of the same fossil records. All these are at odds with the traditional interpretation of these data. The general trend is that average taxic duration is increased. However, the survivorship curves also differ in shape; some appear to be concave up; others seem to be nearly linear or concave down (see fig. 3.5). The shape of such curves has been used to indicate patterns of evolutionary tempo, usually phrased as tachytelic or bradytelic (Simpson 1944). Differences in the amount of concave-downness are apparent in the survivorship curves in figure 3.5. If evolutionary pulse is a function of phyletic splitting and taxic duration, many curve shapes may be concordant with the same fossil evidence, depending on phylogenetic assumptions.

The root of inconsistency between survivorship curves generated from groups and lineages is related to how well the fossil record and the evolutionary process record phylogenetic events (as measured by Z). The discrepancy between lineage and traditional group results is due to a combination of topologic and temporal considerations.

First, if the group with the oldest origin is the sister group to the other

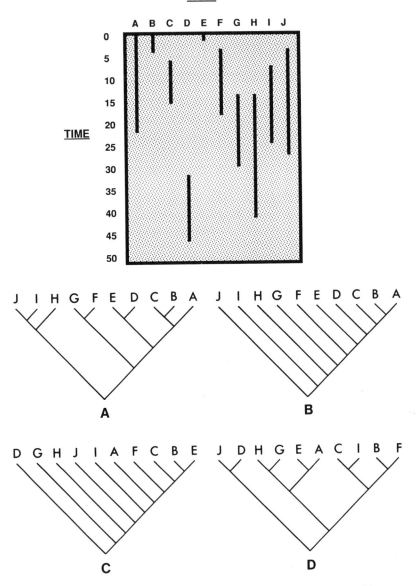

Fig. 3.4. A simulated fossil record for a group containing 10 taxa. Stratigraphic ranges are indicated by solid bars. Four sample phylogenies for the taxa are presented below. Note the variance in correspondence with the fossil record. For example, phylogeny B is grossly discrepant and phylogeny C is strongly correspondent with the record.

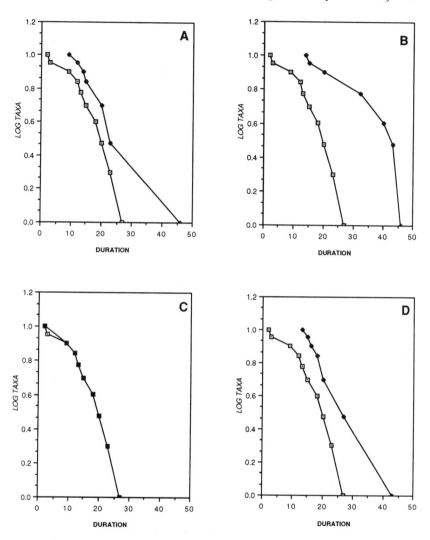

Fig. 3.5. Survivorship curves for groups and phyletic lines extracted from the phylogenies in figure 3.4. Letters refer to the phylogenies in figure 3.4. Open squares represent groups; solid squares represent lineages.

taxa on a "comb" or pectinate (i.e., one where lines split off in a series along a single axis) cladogram, all phyletic lines will have an Al equal to it. This results in survivorship curves that are most unlike those derived from the traditional, Ag, data (see fig. 3.4B). Second, if the order of phyletic splitting roughly approximates the order of appearance of fossils in a "comb" cladogram or if groups of roughly similar ages are each others' sister group, phyletic line survivorship may be similar to group survivorship. In this case Z values are fairly high (see fig. 3.4C). Third, in most cases phylogenies will be a mixture of the preceding factors. The general effect of a phyletic-line approach is the extension of at least some taxic durations because all pairs of sister taxa will probably never appear in the same geological instant. Hence, survivorship curves derived from phylogenetically corrected data will be deflected toward longer durations and will virtually never be identical to those based strictly on the appearance of fossil groups.

Taxonomic rate can be determined accurately from fossil records only if these records are considered to be exact descriptors of evolutionary change and pattern. Reliance on the fossil record as an "exact descriptor" has been questioned by many phylogeneticists (McKenna et al. 1977; Nelson and

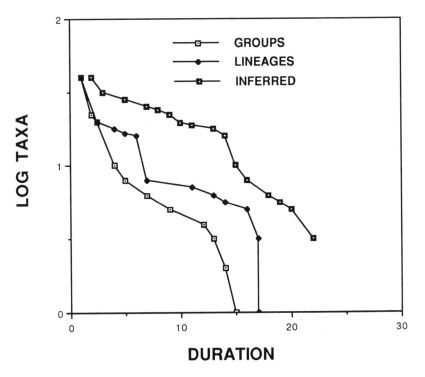

Fig. 3.6. Survivorship curve extracted from data in Szalay and Delson (1979) for 39 genera of primates. Groups derived from fossil occurrences. Lineages are determined from analysis of Szalay and Delson's phylogeny, and are inferred from Szalay and Delson's inference regarding taxic origin.

Platnick 1981). Realizing the shortcomings of uncritical application of the fossil record, Novacek and Norell (1982) generated a survivorship curve for 163 primate genera using the data set of Szalay and Delson (1979). Although Szalay and Delson's tree included nonmonophyletic taxa (ancestral taxa), it is heuristically useful.

In their examination of this curve, Novacek and Norell (1982) noted the difference in curve shape and taxonomic duration between an exact depiction of the fossil record (Dgs) and durations extracted from Szalay and Delson's inferred splitting times for the primate genera. The inferred times were the dashed lines extending back from known fossil ranges illustrated in Szalay and Delson's diagrams. They suggested that the amount of difference between these two results indicates that taxonomic rate, as calculated via a slope function (Van Valen 1973), should be portrayed as a range between these two values. It is difficult to reexamine the entire data set to calculate lineage duration, since many of the taxa are implicitly nonmonophyletic and therefore do not meet one of the requirements for this analysis. In lieu of this I generated three survivorship curves for a sample of the higher primates (platyrhines and catarhines) from Szalay and Delson's data set (fig. 3.6).

Dgs, "inferred durations," and Dls were extracted directly from Szalay and Delson's phylogenies. For the 39 genera analyzed, the result yields a steeply sloped Dg curve, an intermediate Dl curve, and a shallow curve based on durations determined by Szalay and Delson's inference. However, only the Dl curve is a direct prediction that accounts for both the phylogeny and the fossil record.

Example 2: Phylogenetic Hypotheses and Temporal Origination Patterns

Phylogenetic correction affects temporal origination histories in a similar fashion. Such histories are key elements in explaining diversification or demise of phylogenetic groups or in calculating age of groups relative to abiotic events. Calibrated phylogenies can be used to create new origination histories that incorporate phylogeny and depict standing diversity of discrete lineages.

Temporal diversity is usually derived from plots that portray an origination history of a group. These graphs depict number of groups present versus time, where group presence is determined by first and last fossil occurrences. Origination histories and temporal diversities derived from phylogenetic correction of the data in figure 3.4 are contrasted with a group origination history in figure 3.7.

As with survivorship curves, various interpretations are possible, depend-

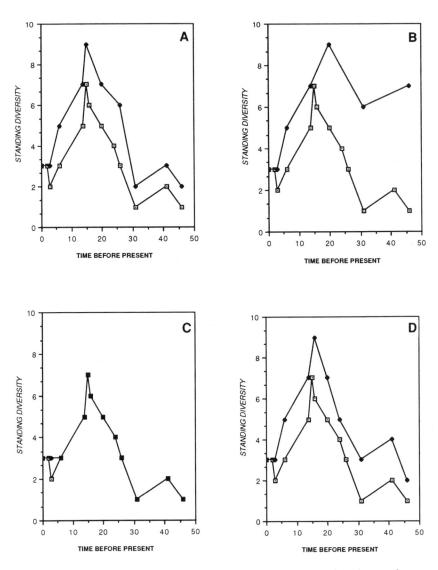

Fig. 3.7. A comparison of standing diversities for groups and phyletic lines utilizing the data set in figure 3.4. Lineage diversity is indicated by the closed boxes and group diversity is indicated by the open boxes. In C, the diversity profiles are superimposed for most of the group's history.

ing on phylogenetic assumptions. Importantly, no single dichotomous phylogeny of the monophyletic taxa in figure 3.4 will yield a lineage origination history identical to the group history determined from the fossil record. It is doubtful that any actual fossil record will ever satisfy the condition where all pairs of sister taxa make first appearances in the same geological instant, resulting in identical temporal distributions of lineages and groups.

Types of incongruence in origination histories between phylogenetic and fossil record approaches reflect general tendencies: (1) Because phylogenetic hypotheses influence only minimum time of origination, extinction "valleys" (intervals of taxonomic decrease) are preserved. However, as pointed out later, they may be masked by increased origination. (2) Peak group diversities determined from the fossil record are generally displaced into the past by lineages. By definition, sister taxa originate at identical times, so the preservation of a single fossil indicates the presence of at least two lineages. (3) Group diversity sets a lower bound for total diversity during any interval. Obviously, the number of monophyletic groups present in the record can never be reduced by a phylogenetic approach. Nevertheless, diversities that are rooted in corrected lineage durations are higher in number than group diversities.

This example is based on only a few phylogenetic cases, yet it demonstrates that phylogenetic correction can severely modify origination diagrams. If individual sister taxa appear in roughly the same temporal instant (Z is high; see fig. 3.7C), the fossil record may, in some cases, be vindicated. However, possibilities of incongruence between the record and phylogenies must also be entertained and expected (see fig. 3.7B).

Diversity diagrams and temporal histories of clades have been important elements in the study of the spatial distribution of organisms and the examination of major events in clade history. An example of each follows. These examples fail to include many complex associated issues, yet they serve heuristically to illustrate the importance of phylogenetic correction of lineage origin.

Many biologists, citing a lack of fossil evidence, suggest that distribution of organisms is the result of dispersal and colonization (Darlington 1957). Common rallying points for such scenarios are the lack of appropriate fossils older than some assumed isolating barrier or the widespread belief that taxa are young (see Nelson and Platnick 1981; Humphries and Parenti 1986). Like the determination of taxonomic rate from a literal reading of the fossil record, this practice fails to consider phylogenetic evidence. When phylogeny is considered, these "fossils" may be found in the form of lineages. For instance, Estes proposed (1983b) that continental rearrangement greatly influenced the contemporary distribution of lizards. This point is somewhat counterintuitive,

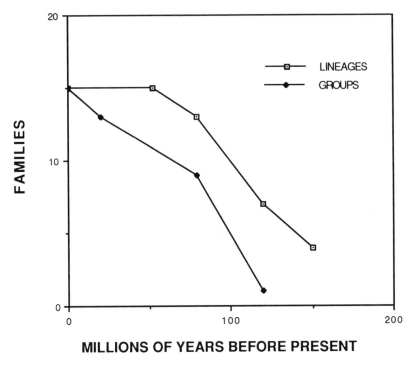

Fig. 3.8. A comparison between lineage and group diversity in the history of extant lizard families. (Data are from Estes 1983.)

since fossils of modern families do not appear in the record until after continental fragmentation (Estes 1983a).

Figure 3.8 plots origination curves for both groups and phyletic lines of extant lizard families taken from Estes' phylogeny (1983a:366). Obviously, splitting events within the lizard clade are older than portrayed by the fossil record. For instance, seven lineages (in addition to two lineages leading to discrete clades not depicted on this plot) were established by the early Cretaceous. Lizards assignable to most modern families do not appear in the fossil record until the latest Cretaceous. Figure 3.8 suggests that phylogenesis of extant lizard families was completed by earliest Tertiary times. This example demonstrates that a phylogenetic treatment may influence opinions about the ages of evolutionary events, and this use of a phylogenetic framework will sharpen hypotheses of biogeographic pattern.

The discrepancies between group and phyletic line origin in Estes' (1983) phylogeny are due to cladistic structure and early occurrence of derived taxa. Among these derived early taxa, the late Jurassic occurrence of the extinct Paramacellodidae, Dorsetisauridae, and Euposauridae as the sister groups to the acrodont iguanoids, the extant Gekkotta, and the Anguidae, respectively, are some of the earliest lizard fossils known. These fossils calibrate four unique lineages that give rise to modern families. The importance of early derived fossils should be emphasized; it points out one of the great advantages of a phylogenetic approach: one fossil can provide information about the history of an entire clade. If these fossils are considered primitive because they are ancient or are tabulated only as single group occurrences, their importance is lost.

The relationships between calibrated phylogenies raises an interesting prospect that will not be considered further here. That is, it may be possible to calibrate phylogenetic trees with temporal estimates of biogeographic fragmentation. In some cases this may allow the historical diversity of groups with particularly poor fossil records to be examined.

Ghost Lineages

Unlike the determination of lineages that corrects the temporal ranges of terminal taxa discussed earlier, the durations of additional unique evolutionary entities can be extracted from a calibrated phylogeny. These additional entities are taxa that are predicted to occur by the internal branching structure of phylogenetic trees. These "additional taxa," however, have become extinct by evolving into new groups. I refer to these as ghost lineages because they are invisible to the fossil record. Ghost lineages can be calibrated and their temporal duration discovered. If ghost lineages are added to the observed terminal lineages discussed earlier, temporal diversities are further inflated and historical patterns of origination may be drastically modified.

Ghost lineages (fig. 3.9, E and F) correspond to ancestral species (Nelson and Platnick 1981) and like other taxa have durations. Ghost lineages are calibrated just like terminal lineages, except that their last occurrence is indicated by their evolution into two daughter lineages rather than extinction (see fig. 3.9). A calibrated phylogeny is shown in figure 3.9. In figure 3.9 this phylogeny is dissected into the minimum number of species present and all lineages are identified. Four of these (A, B, C, D) are observable as fossils or extant taxa, and two (E and F) are ghost lineages. These ghost lineages are the unobservable predictions of the phylogenetic branching of the group ABCD in figure 3.9. In this example, ghost lineage E first appears at time 9,

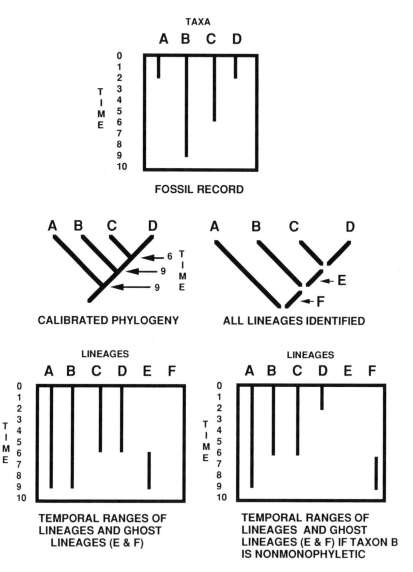

Fig. 3.9. A fossil record and a phylogeny are used to develop a calibrated phylogeny. All lineages are identified and their durations are determined. (See text for details.)

the age of its sister group B. It becomes extinct at time 6 when the first evidence of descendant group CD is indicated by the occurrence of C in the fossil record.

Ghost lineages are important, albeit undetected, elements of diversity. They can be counted like terminal lineages to give the minimum number of discrete taxa predicted to have been present during any time interval. Obviously, some ghost lineages will have a zero length. Zero-duration ghost lineages, such as ghost lineage F in figure 3.9, occur when a calibration is carried down to a lower node because no older fossil is present in the sister group (A in fig. 3.9). If a ghost lineage has a zero duration, its presence is ignored because its presence is included in its descendant lineages. For instance, in figure 3.9, taxon F is not counted because it has "evolved" into B and E. Therefore the first evidence of a zero-length ghost lineage occurrence corresponds with the origination of a descendant group.

In some cases nonmonophyletic species (the metataxa of Donoghue 1985) can be used to demonstrate ghost lineages. For instance, in figure 3.9, if B is not demonstrably monophyletic, it is a putative ancestor until it is excluded from ancestry by the first occurrence of group CD at time 6. Because it remains present in the record it represents a minimum of one lineage. Ghost lineage E, however, now has a duration of zero because it cannot be calibrated by the presence of a nonmonophyletic B. Ghost lineage F acquires a duration. Its lower bound is calibrated by the occurrence of the earliest member of group BCD (here B's occurrence at time 9). Its upper bound is calibrated by the first evidence of splitting, or the origin of group CD. Because the lower part of lineage B does not provide any evidence of lineage splitting (it is a putative ancestor), its first occurrence as well as that of D is specified by the occurrence of C at time 6.

Inclusion of ghost lineages has dramatic affects on perceived evolutionary diversity. Figure 3.10 indicates a fossil record, a cladogram, and a temporal diversity plot for groups, terminal lineages, and all lineages (terminal lineages + ghost lineages). A Z value of 0.9 indicates that the record and cladistically based phylogeny are highly congruent. Terminal lineage and group diversity are similar; however, if ghost lineages are included, actual diversity of unique evolutionary lines is much higher than either terminal lineages or groups.

Temporal diversity patterns can be modified by the addition of ghost lineages to such an extent that they are counterintuitive. For instance, the fossil record in figure 3.11 is combined with the phylogeny. The diversity plot in figure 3.11 displays temporal diversity of groups and all lineages. The graph is self-explanatory. The extinction event suggested by the fossil record at time 3 is a figment of preservation or asymmetrical character evolution if phylogenetic assumptions hold.

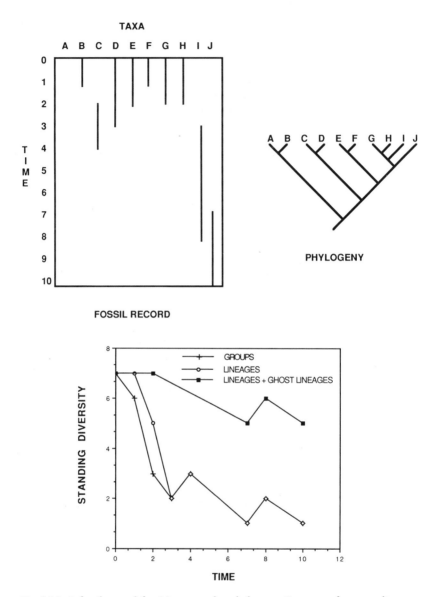

Fig. 3.10. A fossil record for 10 taxa and a phylogeny. Presence of groups, lineages, and ghost lineages was determined and used to construct the diversity profile. Notice the degree of difference caused by the inclusion of ghost lineages, even though lineage diversity is very similar to that of groups.

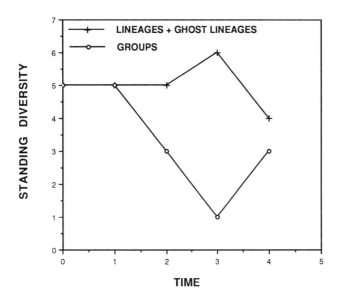

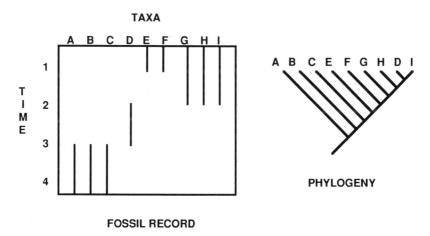

Fig. 3.11. A fossil record, phylogeny and diversity profile for nine taxa. The fossil record appears to indicate taxonomic decrease at time 3, shown on the diversity diagram as the group line. The phylogeny and fossil record corrects these ranges to indicate phylogenetic radiation during this interval.

Lineage diversities are important since they indicate the actual number of evolutionary entities present through time. Lineage diversities elucidate structural levels of organization, by demonstrating the minimum age of hierarchic levels (indicated by sets of derived characters) in the original cladogram.

Like spatial relationships, occurrence of major events influencing evolutionary history can be examined more profitably if temporal diversities are more accurate.

As an example of how new diversity estimates may modify historical explanation, I present the following example. The purported late Cretaceous extinction of Mesozoic communities has been considered by some as the trigger to eutherian mammalian ordinal diversification (Simpson 1978; Stanley 1981; Van Valen 1985b). Traditional interpretations of the fossil record have suggested a rapid demise of Mesozoic taxa, followed by or associated with rapid proliferation and diversification of eutherian mammals (Van Valen 1984, 1985b).

Interpretations of the origin of the eutherian orders suggest that the phylogenetic burst progressed in the Early Paleocene and culminated in the mid-Eocene (Gingerich 1977), a view defended by paucity of Cretaceous eutherian fossils (Archibald 1983; Van Valen 1985a). A departure from this view is that establishment of eutherian diversity is more ancient. This observation tempers reliance on the cause–effect relationship that mammals radiated into divergent groups with the morphologic specializations common to the diverse orders in response to new ecologic opportunities.

Three hypotheses of phylogenetic arrangement of the eutherian orders are applicable (Shoshani 1986; Novacek and Wyss 1986; Miyamoto and Goodman 1986). These hypotheses differ in amount of hierarchic resolution, topology, inclusion of fossil orders, types of data used, and number of orders considered. I calibrated these phylogenies with first fossil occurrence dates in Novacek (1982) as augmented by recent discoveries reported in McKenna (in preparation). A list of these dates is given in table 3.1. Lineage (both lineage and ghost lineage) and group temporal diversity for each phylogeny is presented in figure 3.12. In cases where multiple phylogenetic solutions influence number and age of lineages, I indicate all possible resolutions.

The hypothesis of Novacek and Wyss (1986) considers 18 ordinal-level taxa (fig. 3.12). Origination histories derived from their phylogeny indicate the presence of 9 to 11 components before the end of the Cretaceous. The hypothesis of Shoshani (1986) (fig. 3.12) includes fossil orders and indicates that 14 or 15 lineages were already established by Late Cretaceous times. Only six of the groups considered by Shoshani have been detected in the Late Cretaceous fossil record. Like the Novacek and Wyss phylogeny, that of Miyamoto and Goodman (1986) presents a variety of dichotomous solutions.

Table 3.1. Calibration Dates for the Eutherian Orders 1, 2

Taxon	First Occurrence	Reference
Edentata 4	Middle Paleocene	Scillato Yane 1980; see Marshall et al. 1983
Pholidota	Late Paleocene	Emry 1970
Cetacea	Early Eocene	Gingerich and Russel 1981
Dinocerata	Late Paleocene	Gingerich et al. 1980
Embrithopoda	Late Paleocene	Matthew and Granger 1925
Hyracoidea	Early Eocene	Hartenberger et al. 1985
Proboscidea	Early Eocene	Wells and Gingerich 1983
Desmostylia	Late Oligocene	Ray and McKenna 1986
Sirenia	Early Eocene	Domning 1978
Condylarthra 2,3	Late Cretaceous	de Muizon and Marshall 1987; Sloan and Van Valen 1965
Pantodonta 2	Late Cretaceous	de Muizon and Marshall 1987a
Tubulidentata	Early Miocene	Patterson, 1978
Notoungulata 2	Late Cretaceous	Marshall et al. 1987
Perrissodactyla	Early Eocene	Rose 1981
Artiodactyla	Early Eocene	Rose 1981
Carnivora	Middle Paleocene	Tedford 1976
Primates	Late Cretaceous	Van Valen and Sloan 1965
Scandentia	Late Miocene	Jacobs 1980
Dermoptera	Middle Paleocene	Rose and Simons 1977
Chiroptera	Early Eocene	Van Valen 1979
Lipotyphla	Late Cretaceous	Butler 1972
Macroscelidea	Early Paleocene	Simpson 1931
Lagomorpha	Late Cretaceous	Gregory and Simpson 1926
Rodentia	Late Paleocene	Dawson 1967

1. Occurrences are taken from Novacek (1982) as modified by recent discoveries and McKenna (in preparation).

2. McKenna (personal communication) and Van Valen (1989) question the age of the South American occurrences of these taxa and suggest that they may be early paleocene. This has little bearing on the graphs in figure 3.12 because only Shoshani's includes fossil orders.

3. The monophyly of this taxon is difficult to establish although it was recognized by Shoshani (1986).

4. This taxon may have a latter (Itiboraian) first occurrence (Marshall et al. 1983). This only slightly modifies the conclusions in figure 3.12, because sister taxa are nearly comparable in age.

Still, 10 out of a total 16 calibrated lineages are accounted for in the Late Cretaceous. In all three diagrams all the lineages are accounted for by the early Eocene.

Phylogenetic range correction indicates that early phylogenesis of eutherians occurred in the late Mesozoic and continued on through the Paleogene. All three phylogenies indicate that over 60% of ordinal diversity of eutherians resulted from phylogenesis occurring before the Tertiary. This result contrasts

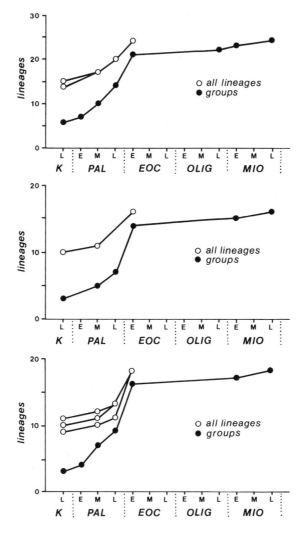

Fig. 3.12. The origination history of the eutherian mammal orders as portrayed by components and groups. (Top) The hypothesis of Shoshani (1986); (middle) the hypothesis of Miyamoto and Goodman (1986); and (bottom) the hypothesis of Novacek and Wyss (1986). The vertical bar indicates the K–T boundary. Calibration information is given in table 3.1.

with relatively low diversities preserved in the fossil record, which in the Late Cretaceous represents less than half of the diversity of evolutionary entities (ghost lineages + lineages).

None of these comparisons between group and line origination falsifies an increase in taxic evolution in the earliest Tertiary. Yet all three phylogenies portray phyletic diversification of Eutheria as occurring earlier than depicted by the paleontological record and also suggest that taxic radiation of the Paleocene was only an extension of a Late Cretaceous phenomenon. The inference collides with standard notions about early eutherian diversification (e.g., Stanley 1979; Van Valen 1985a).

The eutherian example indicates that current phylogenies of the eutherian orders combined with the fossil record offer support for a more ancient diversification of major mammalian groups than is portrayed by the fossil record alone. This result is determined from three independent phylogenies that differ in composition and topology. Yet the pattern of ancient phylogenesis for the eutherian clade persists. This example demonstrates the importance of evaluating diversity histories in reference to specific phylogenetic hypotheses.

Discussion

This approach calls for the injection of phylogenetic analysis into temporally dependent explanations. This approach denies the assertion of instantaneous, dichotomous group formation (see fig. 3.2) as well as the nonphylogenetic compilation of group duration. Instead, it detects taxonomic duration from both the fossil record and phylogenetic hypotheses (see figs. 3.3 and 3.9). In a traditional paleontological approach, group origination is based on actual diagnosable specimens with apomorphies, and a group's duration and first occurrence remain constant for the same sets of specimens.

Phylogenetic assumptions have always been a part of the paleontological research program. These assumptions have only been extended to the recognition of groups (families, genera, etc.) in the record, and their application and usage has been nonuniform. The approach outlined here modifies and gives additional emphasis to phylogenetic assumptions and provides a rationale for their synthesis with the fossil record. In a phylogenetic approach both taxic duration and temporal diversity will be modified by using new phylogenies. The prospect of having phylogenetic realignment influence temporal origination pattern gives the study of evolutionary tempo and mode a hitherto unrecognized test beyond the labors of paleontologists.

A benefit of this approach is its application to groups with a poor fossil record, a problem expected at low taxonomic levels. These comparisons are undoubtedly of the greatest evolutionary interest (Cracraft 1981). The method presented here is based on phylogeny reconstruction, implying that better temporal diversity studies are possible if fossils are combined with phylogenetic hypotheses. This synthesis allows the unrepresented part of clade history (i.e., lineages) to be discovered, and the unknown, paleontologically invisible elements of diversity (ghost lineages) to be tabulated.

Identification of groups in the fossil record and their tabulation into origination histories is only an initial step in the determination of evolutionary tempo, mode, and pattern. Although simple tabulation of taxa, if monophyletic, provides an absolute minimum bound on diversity and longevity, inferring patterns from these compilations is doomed to failure, because pattern of preservation is usually inconsistent with any phylogeny of monophyletic taxa, and ghost lineages are paleontologically invisible. A better approach is to make the assumption of phylogeny explicit and to examine these questions within the phylogenetic system. Thereby, the predictions of specific phylogenetic hypotheses can be extended to the temporal dimension.

The following list summarizes the major points of the above argument:

1. Evolutionary tempos and evolutionary patterns should reflect phylogenesis rather than preservation.
2. If phyletic splitting and monophyly are constraints on group membership, phylogenetic assumptions are inherent in any compilation of taxonomic diversity and duration.
3. Naive compilation of fossil durations through time gives origination histories and taxonomic rates incompatible with patterns derived from any possible phylogeny of monophyletic taxa.
4. Study of temporal pattern and evolutionary rate can be more rigorously accomplished if the assumption of phylogeny and the measurement of phyletic longevity is retrieved from cladistic structure integrated with the fossil record.

ACKNOWLEDGMENTS

I thank Tim Collins, Michael Donoghue, Kevin de Queiroz, Malcolm McKenna, Peter Meylan, and Paul Sereno for discussing ideas presented here. Joel Cracraft, John Flynn, John Gatesy, Gary Nelson, Michael Novacek, Keith Thomson, Bruce Tiffney, Jim Valentine, Elisabeth Vrba, Quentin Wheeler, and Andy Wyss read this or an earlier draft of this manuscript and their comments are greatly appreciated. Mr. Joseph Herron is thanked for his support of this and other projects. The Department of Vertebrate Paleontology, American Museum of Natural History, and the Department of Biology, Yale University, provided support.

REFERENCES

Archibald, J. D. 1983. Structure of the K-T mammal radiation in North America. *Acta Paleont. Polon.* 28:7–17.

Brooks, D. R. and E. Wiley. 1986. *Evolution as Entropy: Toward a Unified Theory of Biology.* Chicago: University of Chicago Press.

Butler, P. M. 1972. The problem of insectivore classification. In A. Joysey and T. S. Kemp, eds., *Studies in Vertebrate Evolution,* pp. 253–265. New York: Winchester Press.

Cracraft, J. 1981. Pattern and process in paleobiology: The role of cladistic analysis in systematic paleontology. *Paleobiology* 7:456–468.

Cracraft, J. 1985. Conceptual and methodological aspects of the study of evolutionary rates, with comments on bradytely in birds. In N. Eldredge and S. M. Stanley, eds., *Living Fossils,* pp. 95–104. New York: Springer Verlag.

Darlington, P. J. 1957. *Zoogeography: The Geographic Distribution of Animals.* New York: John Wiley.

Dawson, M. R. 1967. The fossil history of the families of recent mammals. In S. Anderson and J. Knox Jones, Jr., eds., *Recent Mammals of the World: A Synopsis of Families,* pp. 12–53. New York: Ronald Press.

de Muizon, C. and L. G. Marshall. 1987a. Le plus ancien Pantodonte (Mammalia) du crétacé supérieur de Bolivie. *Compt. Rend. Hebd. Acad. Sci. Paris* 304:205–208.

de Muizon, C. and L. G. Marshall. 1987b. Le plus ancien Condylarthre (Mammalia) sud américain (Crétacé supérieur, Bolivie). *Compt. Rend. Hebd. Acad. Sci. Paris* 304:771–774.

Dingus, L. and P. M. Sadler. 1982. The effects of stratigraphic completeness on estimates of evolutionary rates. *Syst. Zool* 31(4):400–412.

Domning, D. P. 1978. Sirenia. In V. J. Maglio and H. B. S. Cooke, eds., *Evolution of African Mammals,* pp. 573–581. Cambridge: Harvard University Press.

Domning, D. P., C. E. Ray, and M. C. McKenna. 1986. Two new Oligocene Desmostylians and a discussion of tethytherian systematics. *Smithson. Contrib. Paleontol.* 59:1–56.

Donoghue, M. J. 1985. A critique of the biological species concept and recommendations for a phylogenetic alternative. *The Bryologist* 88(3):172–181.

Eldredge, N. and S. J. Gould. 1972. Punctuated equilibria, an alternative to phyletic gradualism. In T. Schopf, ed., *Models in Paleobiology,* pp. 82–115. San Francisco: W. H. Freeman.

Eldredge, N. and J. Cracraft. 1980. *Phylogenetic Patterns and the Evolutionary Process.* New York: Columbia University Press.

Eldredge, N. and M. J. Novacek. 1985. Systematics and paleontology. *Paleobiology* 11(1):65–74.

Emry, R. J. 1970. A North American Oligocene pangolin and other additions to the Pholidota. *Amer. Mus. Nat. Hist. Bull.* 142(6):455–510.

Estes, R. 1983a. Sauria terrestria, Amphisbaenia. *Handbuch der Paleoherpetologie Part 10A.* Stuttgart: Gustav Fischer.

Estes, R. 1983b. The fossil record and the early distribution of lizards. In A. G. J.

Rhodin and K. Miyata, eds., *Advances in Herpetology and Evolutionary Biology: Essays in Honor of E. E. Williams*, pp. 365–398. Cambridge: Harvard University Press.

Felsenstein, J. 1978. The number of evolutionary trees. *Syst. Zool.* 27(1):27–33.

Gauthier, J. A. 1984. *A cladistic analysis of the higher systematic categories of the Diapsida.* Dissertation. Berkeley: University of California.

Gingerich, P. D. 1977. Patterns of evolution in the mammalian fossil record. In A. Hallam, ed., *Patterns of Evolution.* Amsterdam: Elsevier.

Gingerich, P. D., K. D. Rose, and D. W. Krause. 1980. Early Cenozoic mammalian faunas of the Clark's Fork Basin–Polecat Bench Area, northwestern Wyoming. In P. D. Gingerich, ed., *Early Cenozoic Paleontology and Stratigraphy of the Bighorn Basin, Wyoming.* Papers in Paleontology, Univ. Mich. no. 24, 51–68.

Gingerich, P. D. and D. E. Russell. 1981. *Pakicetus inarchus* a new archaeocete (Mammalia, Cetacea) from the early-middle Eocene Kuldana Formation of Kuhat (Pakistan). *Contrib. Mus. Paleontol. Univ. Mich.* 25(11):235–246.

Gregory, W. K. and G. G. Simpson. 1926. Cretaceous mammal skulls from Mongolia. *Amer. Mus. Nat. Hist. Novit.* 225:1–20.

Hartenberger, J.-L., C. Martinez, A. Ben-Saïd. 1985. Découverte de Mamifères d'âge Éocène inférieur en Tunisie Centrale. *Compt. Rend. Hebd. Acad. Sci. Paris* 301:649–652.

Hennig, W. 1965. Phylogenetic systematics. *Ann. Rev. Entomol.* 10:97–116.

Humphries, C. and L. Parenti. 1986. *Cladistic Biogeography.* New York: Oxford University Press.

Jacobs, L. L. 1980. Siwalik fossil tree shrews. In W. P. Luckett, ed., *Comparative Biology and Evolutionary Relationships of the Tree Shrews*, pp. 205–216. New York: Plenum Press.

Kurtén, B. 1960. Chronology and faunal evolution of the earlier European glaciations. *Soc. Sci. Fenn. Comm. Biol.* 21:40–62.

Levinton, J. S. and J. S. Farris. 1987. On the estimation of taxonomic longevity from Lyellian curves. *Paleobiology* 13(4):479–483.

Marshall, L. G., R. Hoffstetter, and R. Pascual. 1983. Mammals and stratigraphy: Geochronology of the continental mammal bearing Tertiary of South America. *Palaeovertebrata Mem. Extraord.* 93 pp.

Marshall, L. G., C. de Muizon, and B. Sigé. 1987. *Perutherium altiplanensis*: un notongulé du Crétacé supérieur du Pérou. *Palaeovertebrata* 13:145–155.

Matthew, W. D. and W. Granger. 1925. Fauna and correlation of the Gashato Formation of Mongolia. *Amer. Mus. Nat. Hist. Novit.* 186:1–12.

McKenna, M. C., G. F. Engelman, and S. F. Barghorn. 1977. Review: "Cranial anatomy and evolution of early Tertiary Plesiadapidae (Mammalia, primates)" by P. D. Gingerich. *Syst. Zool.* 26:233–238.

Miyamoto, M. M. and M. Goodman. 1986. Biomolecular systematics of eutherian mammals: Phylogenetic patterns and classification. *Syst. Zool.* 35(2):230–240.

Nelson, G. and N. Platnick. 1981. *Systematics and Biogeography: Cladistics and Vicariance.* New York: Columbia University Press.

Norell, M. A. 1989. Late Cenozoic lizards of the Anza Borrego Desert, California. *Los Angeles County Mus. Contrib. Sci.* 414:1–31.

Novacek, M. J. 1982. Information for molecular studies from anatomy and fossil evidence on higher eutherian phylogeny. In M. Goodman, ed., *Macromolecular Sequences in Systematic and Evolutionary Biology,* pp. 3–41. New York: Plenum Press.

Novacek, M. J. and M. A. Norell. 1982. Fossils, phylogeny, and taxonomic rates of evolution. *Syst. Zool.* 31(4):366–375.

Novacek, M. J. and A. R. Wyss. 1986. Higher level relationships of the eutherian orders: Morphologic evidence. *Cladistics* 2(3):257–287.

Patterson, B. 1978. Pholidota and Tubulidentata. In V. J. Maglio and H. B. S. Cooke, eds., *Evolution of African Mammals,* pp. 268–278. Cambridge: Harvard University Press.

Raup, D. M. and G. E. Boyajian. 1988. Patterns of generic extinction in the fossil record. *Paleobiology* 14(2):109–125.

Raup, D. M. and D. Jablonski. 1986. *Patterns and Processes in the History of Life.* Berlin: Springer-Verlag.

Raup, D. M. and L. G. Marshall. 1980. Variation between groups in evolutionary rates: A statistical test of significance. *Paleobiology* 6:9–23.

Rose, K. D. 1981. The Clarkforkian land-mammal age and mammalian faunal composition across the Paleocene–Eocene boundary. *Pap. Paleontol. Univ. Mich.* 26.

Rose, K. D. and E. L. Simons. 1977. Dental function in the Plagiomenidae: Origin and relationships of the mammalian order Dermoptera. *Contrib. Mus. Paleontol. Univ. Mich.* 24:221–236.

Rowe, T. 1987. Definition and diagnosis in the phylogenetic system. *Syst. Zool.* 36(2):208–211.

Schindel, D. 1980. Microstratigraphic sampling and the limits of paleontological resolution. *Paleobiology* 6:408–426.

Scillato Yane, G. J. 1980. Nuevo genero de Dasypodidae (Edentata: Xenarthra) del Plioceno de Catamarca (Argentina); algunas consideraciones filogeneticas y zoogeograficas sobre los Euphractini. *Cong. Argentine Paleont. Bioestrat. Actas* 1(2):4449–4461.

Sepkoski, J. J., Jr. 1988. Alpha, beta, or gamma: Where does all the diversity go? *Paleobiology* 4(3):221–234.

Shoshani, J. 1986. Mammalian phylogeny: A comparison of morphologic and molecular results. *Molec. Biol. Evol.* 3(3):222–242.

Simpson, G. G. 1931. A new insectivore from the Oligocene Ulan Gochu Horizon, of Mongolia. *Amer. Mus. Nat. Hist. Novit.* 505:1–22.

Simpson, G. G. 1944. *Tempo and Mode in Evolution.* New York: Columbia University Press.

Simpson, G. G. 1953. *The Major Features of Evolution.* New York: Columbia University Press.

Simpson, G. G. 1978. Early mammals in South America: Fact, controversy and mystery. *Proc. Amer. Phil. Soc.* 122:318–328.

Sloan, R. B. and L. Van Valen. 1965. Cretaceous mammals from Montana. *Science* 148:220–227.

Stanley, S. M. 1979. *Macroevolution: Patterns and Process.* San Francisco: W. H. Freeman.

Stanley, S. M. 1981. *The New Evolutionary Timetable.* New York: Basic Books.

Szalay, F. S. and E. Delson. 1979. *Evolutionary History of the Primates.* New York: Academic Press.

Tedford, R. H. 1976. Relationship of pinnipeds to other carnivores (Mammalia). *Syst. Zool.* 25:363–374.

Van Valen, L. 1973. A new evolutionary law. *Evol. Theory* 1(1):1–31.

Van Valen, L. 1979. The evolution of bats. *Evol. Theory* 4:129–142.

Van Valen, L. 1984. Catastrophes, expectations and evidence. *Paleobiology* 10:121–137.

Van Valen, L. 1985a. Why and how do mammals evolve unusually rapidly? *Evol. Theory* 7(3):127–132.

Van Valen, L. 1985b. A theory of origination and extinction. *Evol. Theory* 7(3):133–142.

Van Valen, L. and R. E. Sloan. 1965. The earliest primates. *Science* 150:743–745.

Vrba, E. S. 1985. African Bovidae: Evolutionary events since the Miocene. *South African J. Sci.* 81:263–266.

Wells, N. A. and P. D. Gingerich. 1983. Review of Eocene Anthracobunidae (Mammalia: Proboscidea) with a new genus and species, *Jozartia palustris,* from the Kuldana Formation of KohSat (Pakistan). *Contrib. Mus. Paleontol. Univ. Mich.* 26(7):117–139.

4 : Extinction and the Origin of Species

Kevin C. Nixon and Quentin D. Wheeler

In a former book, in which some of the traditional and
chronic controversies of biology were studied, it was
shown how largely these disputes were traceable either to
failure to eliminate metaphysical elements from biological
topics, or to difficulties created by the shortcomings of
current biological language.　　—Woodger (1937: viii)

Abstract. Given a phylogenetic species concept, extinction of plesiomorphic character
states results in speciation, which we equate with character transformation. Species
are the end products of evolution, marked by unique character distributions, and do
not have internal phylogenetic structure. These unique character distributions arise
through extinction of plesiomorphic states and simultaneous fixation of apomorphic
states. From a cladistic perspective, such changes are irreversible and provide evidence of historical patterns. In contrast, microevolutionary changes that involve only
variably distributed attributes among tokogenetic elements do not have the qualities
of characters and cannot be used to infer phylogenetic pattern. Apparent anagenesis
can be an artifact of extinction, in that the species bearing "transitional" character
states (adjacent states) have disappeared through extinction, leaving cladistic branches
marked by numerous character steps. It is important to separate "methodological"
anagenesis from discussions of rates of evolution in nature. Methodologically, character transformations occur only at the time of species genesis, and only arbitrary or
process-based definitions of species recognize anagenetic change (involving fixation)
within species. The term *cladogenesis* is sufficient and correct for species genesis (each
species is a new clade) and is comparable to, and marked by, character transformation. In sexually reproducing species there exists a correlation between rates of lineage extinction and species declination, diversification, and morphological "disparity." In asexual forms, lineage extinction is correlated only with species declination.

In recent literature, the term *extinction* is widely used but rarely given precise definition. It is clear that in its most general usage, extinction refers to the loss, or "death," of whole species (Raup 1984) or entire isolated populations of a species (Diamond 1984). We would like to extend our discussion of extinction to the death of lineages within sexual populations and focus on the implications of such extinctions in terms of cladistic theory. Our premise is that intrapopulational extinction of lineages is an "organizing" factor that results in the nested hierarchy of characters that is detectable through cladistic analyses. Such a model is not a prerequisite for cladistic analysis, but instead is offered as an explanation of the hierarchic resolution detectable with cladistic analysis. The processes of mutation, recombination, death, and lineage extinction are directly observable in the field or laboratory, thus contributing to the robustness of this explanation. The acceptance of such an explanation relegates particular models of process (e.g., selection and speciation) to the level of secondary explanations, which are not necessary to explain the hierarchic diversity of life (see also Vrba and Gould 1986). Whether patterns of lineage extinction are random or ordered (such as some process of selection), species and monophyletic groups (sets of species bearing unique characters) will exist in nature as a result of mutation, recombination, dispersal, and extinction.

Hierarchy and Anarchy

Intrinsically hierarchic as well as nonhierarchic patterns of relationship occur among living things. Distinguishing between these different patterns is relevant to the selection of appropriate objects (*elements* in set terminology) and methods of logical analysis. In particular, organisms, populations, and species are of interest to systematists, and different assumptions and methods of analysis must be applied, depending on which level is under study. When the elements under consideration are species, cladistic analysis can be undertaken to resolve a pattern of hierarchically nested characters that include these elements, and phylogenetic interpretations of such nested patterns can be made (Hennig 1966). The preferred method of cladistic analysis utilizes global parsimony (e.g., Farris 1979, 1982, 1983) and assumes that a nested hierarchic pattern occurs in the distribution of characters among the elements (Nelson and Platnick 1981). The application of parsimony to resolve hierarchic structure in a set of elements assumes that homoplasy is due to error in assignment of character homology. When the pattern of characters among the elements is not hierarchic, the assumptions of cladistics are violated and cladistic analysis using global parsimony cannot resolve intrinsic hierarchies,

but may result instead in hierarchic structure that does not reflect phylogenetic pattern in the sense of Hennig (1966). Such a situation occurs to varying
degrees in systems of birth relationship among individuals, called tokogenetic
systems by Hennig (1966). In sexual (e.g., biparental) tokogenetic systems,
because of recombination, the character hierarchy that might be discovered
does not necessarily reflect an intrinsic pattern of relationship among the
elements (individuals in this case), nor does it necessarily reflect an intrinsic
sequence of character transformation. Elements (individuals) in such a nonhierarchic system are most accurately grouped into subsets with incomplete
intersections (fig. 4.1), because recombination and segregation break down
the original hierarchic character pattern inherent in the mutational process. If
such systems are forced into strictly nested hierarchies, some homologous

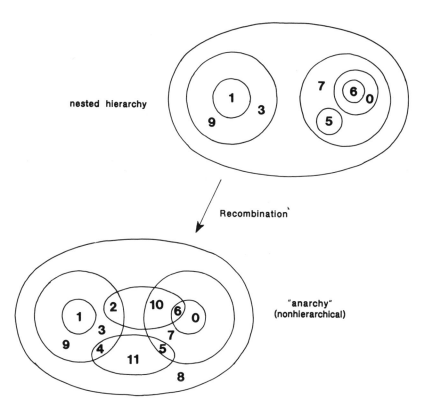

Fig. 4.1. Set representation of hierarchic and nonhierarchic patterns. Mutation, producing hierarchic pattern, is the source of attributes that become distributed in an
incompletely intersecting pattern due to recombination.

attributes (attributes of common origin) in different individuals must be interpreted as homoplasy. Homoplasy under these circumstances does not reflect error in homology assignment; instead it reflects a basic violation of the assumption of hierarchic relationship of elements, and the application of parsimony to reduce homoplasy is not justified. In asexual organisms, tokogenetic patterns occur within clones (species in a strict sense), where the pattern of birth relationship is not detectable because no character variation exists on which to base nested sets. Our definition of species (Nixon and Wheeler 1990) differs from that of Hennig (1966), who in essence did not recognize species in asexual organisms. Hence we will refer to tokogenetic patterns in both sexual and asexual taxa as nonhierarchic. Although birth relationships in completely asexual organisms are trivially hierarchic, such patterns are not detectable within genetically identical clones except by direct observation of birth. Various degrees of hierarchic and nonhierarchic pattern occur within and between sexual populations, but cladistic analysis does not test hierarchic structure. Our proposition is that tokogenetic systems do not meet the assumptions of cladistic analysis and cannot be considered to be fully hierarchic and are therefore inappropriate for cladistic investigations. The question of "robustness" of cladistics in nonhierarchic systems cannot be addressed here.

With the widespread acceptance of cladistic methodologies, homology has come to be equated more or less directly with synapomorphy and the concept of characters as hierarchically nested attributes (Patterson 1982:65): "If homology is the property of monophyletic groups, several consequences follow. First, homology and synapomorphy must be the same thing. Second, homologies must form a hierarchy. And third, nonmonophyletic groups (para- and polyphyletic) cannot be characterized by homology." Clearly, in a phylogenetic or nested hierarchic system, homology is roughly equivalent to synapomorphy, and as such is equivalent to the nested character concept where a set or subset includes all elements that bear it (Nelson and Platnick 1981). However, in a nonhierarchic system synapomorphy and homology are not equivalent, because the concepts of synapomorphy and monophyly are not applicable in nonhierarchic systems (Nixon and Wheeler 1990; Nixon 1991). In such a system, homology must be simply "shared attribute of common origin." Characters in the hierarchic sense cannot exist in a nonhierarchic system. Therefore they are neither apomorphic nor plesiomorphic in a cladistic sense and can be viewed as such only in a trivial sense of time of origin. We prefer to call attributes such as variably distributed alleles within sexual populations *traits*. Species exist as "kinds" of things or elements suitable for cladistic analysis regardless of whether they are diagnosable with apomorphic characters or only unique combinations of plesiomorphic characters (Nixon and Wheeler 1990). Therefore the attributes by which species are identified must be designated independently of polarity and we see no alternative but to differentiate between character and character state (see also Pimentel and Riggins, 1987). A character *state* under such a definition is not an inclusive character and might mark a monophyletic group of species, a paraphyletic group of species, a single "apomorphic" species, or a single "plesiomorphic" species; or it might consist of nonhomologous components. In contrast, characters in the sense of Platnick (1979) are concepts that are applied to nested hierarchies (e.g., cladograms). We restrict the use of *character state* to species diagnosis and aspects of character analysis that must be performed prior to understanding polarity of characters (e.g., in the development of a character state matrix for cladistic analysis). Following character analysis, groupings of species are determined through cladistic analysis (or ontogenetic analysis, if applicable; see Nelson 1978; Wheeler 1990a, 1990b); therefore the pattern of nested characters can be discovered. A large portion of characters in many studies can be resolved only through the analysis and ordering of character states with the use of an outgroup (Watrous and Wheeler 1981; Wheeler 1981). Character state is therefore a generalized hypothesis of homology that lacks an explicit statement of polarity relative to characters that are more generalized or are modifications of it.

Extinction can be generalized from an evolutionary concept to an analogous concept in set theory and cladistic analysis in its broadest sense. Extinction as a purely theoretical concept can be defined as the removal of a unique element from the universal set of elements. In cladistic analysis, this universal set of elements will typically be a set of extant and extinct species or species groups that are treated as terminal units or monophyletic groups. We must distinguish between the removal of an element from *consideration* in an analysis and the removal of an element from *participation* in evolution. Extinct forms that are known as fossils have been removed from evolution and cannot affect future patterns of diversity, although they may still be available for analysis. The removal of unique elements will have different effects on cladistic analysis, depending on whether the patterns of trait or character distribution are initially hierarchic or nonhierarchic. Because of this, our discussion of extinction must begin with a discussion of species. Our basic premise is that in sexual organisms, species exist as the smallest units (elements) that have an intrinsically hierarchic pattern of character distribution among them (Nixon and Wheeler 1990). Another way of viewing this is to realize that species are the largest groupings of organisms that do not have internal phylogenetic structure. In general, groups of species have nested hierarchic patterns of relationship based on character distributions, whereas groups of individuals within sexual species do not. To develop these ideas, it is necessary to discuss in more detail species concepts as they relate to cladistics.

Since Darwin (1859), evolution often has been defined as "descent with modification." Were this literally true, we would live in a world occupied by one massive polymorphic species, not unlike the genetic continuum envisioned by early architects of the New Synthesis. Evolutionary change is marked by a series of modifications and extinction events. Speciation occurs when the extinction of plesiomorphic analogs is completed. Until this point there is a single polymorphic species. Without the removal of plesiomorphic analogs, either completely or from local populations, there are no character transformations or new species. We therefore define macroevolution as descent with modification (attributable to mutation and/or recombination) and extinction. In contrast, we regard microevolution as the various infraspecific phenomena involving traits, such as the change in frequency of traits within populations.

Considerable confusion regarding the term *monophyly* has been introduced recently through a proposed extension of the concept of phylogenetic systematics to populations, individuals, and cell lines, among other things (de Queiroz and Donoghue 1990). This new version of phylogenetic systematics is based on a vague definition of *common descent* that is independent of hierarchy and character transformation (e.g., *descent with modification* is too restrictive). Thus de Queiroz and Donoghue apply the term *monophyletic* to groups of elements that do not have hierarchic patterns of relationship, such as all descendants of a particular individual within the sexual population. If such terminology is widely adopted, the term *monophyletic* will be imprecise and impossible to use without extensive explanation (Nixon, in press 1991). We therefore restrict our use of *monophyletic* to the sense in which Hennig (1966) used it, to refer to an ancestral species and all species, extant or extinct, descended from it. As discussed here and elsewhere, the term *lineage* is widely used and is adequate and sufficient for a broad concept of descent that includes both hierarchic and nonhierarchic patterns (Nixon, in press 1991).

Species Concepts

Species concepts provide the "units" or elements that are ordered by cladistic analysis into nested sets based on character distributions. Such nested hierarchies can be interpreted as reflective of evolutionary history, and as such are of great interest to most biologists. Because species are the ultimate elements of cladistic analysis, species concepts will affect our view of organic diversity and our ability to reconstruct the history implicit in that diversity. For

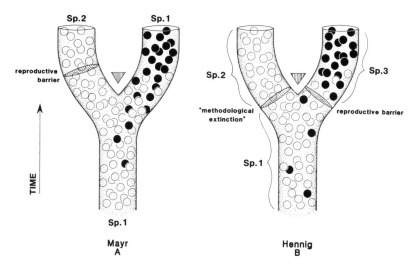

Fig. 4.2. Graphic representation of Mayrian and Hennigian species concepts. (A) Mayrian concept, in which species are delimited by reproductive barriers. (B) Hennigian species concept, in which species are delimited by divergence events and are therefore "internodal" in nature.

phylogenetic inference, the goal of the systematist is to order the elements, or species, so that the system of ordering most closely reflects phylogenetic history. Therefore the systematist needs to adopt a species concept that is consistent with this goal.

The "biological" species concept (Mayr 1942, 1963) restricts our view of evolution and does not provide elements that can be analyzed consistently with cladistic methods. Under a Mayrian* species concept, evolutionary history can be viewed as a branched tube of reproductive relationship, with genetic change occurring within the constraints of this system (fig. 4.2A). The Mayrian species concept in its strictest form, as practiced by evolutionary biologists, produces an arbitrary view of species through time. One of the greatest flaws of such a concept is that individuals within any "species" are connected by reproduction back into the past with all ancestors and through these ancestors to all other species. Therefore using reproductive potential as a criterion for species boundaries is logically inconsistent. Species exist only as relationships among populations at a particular moment, and the use of "potentials" to define entities results in unavoidable subjectivism. Within a

---

*We prefer to use the term *Mayrian* for what has been termed the biological species concept and has been separated into either the IC (isolation concept) or RC (recognition concept) by some authors (Paterson and Macnamara 1984; Paterson 1985).

given time span for Mayrian species, the same populations can be considered to be one or many species, and some criterion based on amount of character change must be applied to distinguish among related populations through time. Wright (1982:437) suggested that if "there have been periods of apparent stasis, separated by apparently abrupt, or at least very rapid, change, the boundaries are most conveniently made at the latter times." Thus Mayrian species are arbitrary, or at best ambiguous, when viewed temporally. Such arbitrary units as Mayrian species cannot be used for phylogenetic analyses without certain modifications of the concept.

Hennig (1966) modified a Mayrian species concept to make it more consistent with his view of phylogenetic analysis. Hennig realized that reproductive isolation alone was not sufficient to view species in a historical perspective. Instead of an arbitrary view of species boundaries through time, Hennig viewed every cladogenetic event as producing two new species, with the loss of the ancestral species through extinction, whether or not identifiable character changes occurred in both descendent species. This constraint produced species that were equivalent to the internodes of a cladogram (fig. 4.2B). The Hennigian species concept was therefore consistent and could be used in a historical context, whereas the Mayrian species concept could not. However, the constraint of ancestral extinction necessary to adjust Mayrian species to cladograms means that evolutionarily insignificant "events" mark the end, or extinction, of species, and two populations viewed as separate species could be indentical for all characters. This notion of "automatic" ancestral extinction and dichotomous speciation has received little support as a biological phenomenon, with the exception of Brundin (1972) and Ridley (1989).

Both the Mayrian and Hennigian species concepts allow anagenetic character change within species through time (fig. 4.3; see also Hennig 1966: fig. 14). As pointed out earlier, the only way to apply a Mayrian species concept to paleontological samples is to delimit species arbitrarily on the basis of amount or abruptness of change (Wright 1982). In contrast, the Hennigian species concept allows nonarbitrary estimates of both numbers of species and levels of anagenetic change, if a phylogeny has been hypothesized.

The Hennigian species concept retains numerous flaws, even though it resolved certain problems traceable to its Mayrian basis. We view the Hennigian species concept as arbitrary because otherwise identical populations are considered to be different species at different points on the cladogram, on the basis of patterns that occcur independently in other species. The shifting nature of species delimitation found in a Hennigian species concept is also found in autapomorphic species concepts, where species cease to be species because certain populations become fixed for autapomorphic states. Such

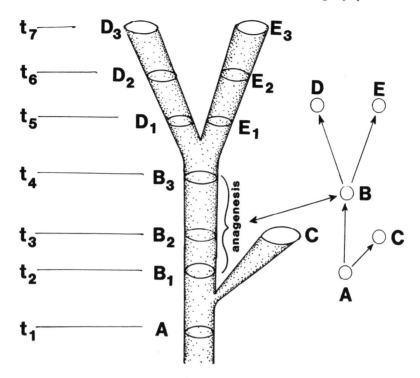

Fig. 4.3. Graphic representation of Hennigian species concept, in which character changes within species that do not result in divergence or vicariance are viewed as anagenetic change within a single species. Modified from Hennig (1966: fig. 14).

methodological constraints do not allow us to view species as elemental components of phylogenetic history, although it is clear that Hennig viewed species as such: "The species are, in the sense of class theory, the elements of the phylogenetic relationship" (Hennig 1966:29).

Phylogenetic Species Concepts

In its broadest sense, a phylogenetic species concept is traceable to Rosen (1978, 1979), who proposed that species are the smallest diagnosable autapomorphic units amenable to cladistic analysis. Some later authors have promoted this concept, referring to it as a "monophyletic" species concept (de Queiroz and Donoghue 1988), but it is more accurately designated an "autapomorphic" species concept, because of the inapplicability of the concept of monophyly at the species level (Nixon and Wheeler 1990; Nixon 1991). Other authors, however, defined species as the smallest diagnosable clusters of individuals, without reference to autapomorphies (e.g., Eldredge and Cracraft 1980; Cracraft 1983; Nelson and Platnick 1981). Nixon and Wheeler (1990) further explicated a phylogenetic species concept in which species must be diagnosable by constantly distributed unique combinations of characters. Character constancy in a population or series of populations is the result of extinction of alternative states. Thus, put simply, there is a correspondence between *trait extinction* and *character transformation.*

Among the types of phylogenetic species concepts outlined earlier, the autapomorphic species concept (ASC) can only be implemented in a static time frame. Autapomorphic species cease to be species and will become "metaspecies" when speciation events (character transformations) occur that make them ancestral, or "paraphyletic" in the sense of de Queiroz and ·Donoghue (1988; see also Kluge 1989; Nelson 1989b). This is not the case with the other approach to phylogenetic species, where species are diagnosed by intrinsic character information, without hypotheses about relationships either above or below the species level. Autapomorphically diagnosed species cannot be used to describe the internal nodes of cladograms, thus implying that there are no ancestral species, only descendant species. In addition, autapomorphic species usually represent only a portion of the terminal lineages of a cladogram, although with some data sets all terminal lineages are marked by autapomorphies and can be assigned species status.

Species are uniquely different from higher-level taxa, in that species do not have resolvable internal phylogenetic structure among the individual organisms included in the species (Hennig 1966; Nixon and Wheeler, 1990). We disagree with Nelson (1989a; 1989b:287) that, "A species is only a taxon." The discovery of completely resolvable phylogenetic structure among all individuals, groups of individuals, or populations within what we believe to be a single species falsifies our hypothesis that the more inclusive group is a species. Species are groupings of individual organisms. Higher monophyletic taxa are groupings of species. For this reason, when discussing species in a general sense, we refer to them as elements, which is consistent with terminology commonly used in set theory (Quine 1961). Because of their elemental (cladistically indivisible) nature, the discovery and diagnosis of species is not a ranking or a grouping procedure. Instead, species are the smallest phylogenetic elements not further *divisible* into phylogenetically related elements, whereas higher taxa are groupings of elements into subsets (clades) based on the distribution of hierarchically nested characters.

If the discovery of relationships among species is the goal of phylogenetic analysis, as repeatedly emphasized by Hennig (1966, e.g., pp. 206–209; 213:

"basic process of phylogeny, which is species cleavage"), then *species* are the *elements* of phylogenetic (hierarchic) analysis. In contrast, *individuals* are the *elements* of tokogenetic (birth relationship) analysis. Sets of species that bear a homologous character are considered to be monophyletic, relative to species not bearing the character, based on the assumption that their relationship is hierarchic. Because of a lack of hierarchic structure within sexual populations, the same cannot be said for sets of individuals that bear a homologous trait within species. An autapomorphy is simply a character (or trait) that, when viewed as a set, includes a single element, or species. The inclusion of an element in a particular set does not change it from being an element to being a nonelement; indeed, the element must be determined to belong to the universal set of elements (all species) before it is placed in any particular character set. Determination that a species has an autapomorphy has no bearing on the tokogenetic relationships of its component individuals. Indeed, it would seem impossible, without some specific model of evolution, to say that the individuals within one species are "monophyletic" while the individuals within another species are not. If "monophylesis" of populations within species is dependent on time of divergence, which seems likely, then any two sister species are *equally* likely to be "monophyletic" because they diverged from each other simultaneously.

Clades are sets of species diagnosable by the distribution of one or more homologous characters. Monophyletic groups are equivalent to clades, but by definition monophyletic groups include more than one species, because monophylesis is a statement of relationship among elements. Given a phylogenetic species concept, speciation is equivalent to cladogenesis, and a clade including a single species is the smallest possible clade. Once again, in this case, the species (element) is not equivalent to the clade (subset) to which it belongs. Clades exist independently of whether true "branching" occurs or whether, in the process of character transformation, the ancestral species becomes extinct. In either case, character transformation results from the origin of a modification or modifications (mutation) followed by the extinction of the plesiomorphic state(s), either globally or locally.

The phylogenetic species concept provides opportunities for a reevaluation of the concept of character. In recent years, discussion of characters has centered on the concept of homology. The first step in character construction must be a hypothesis of homology. Homology, in the phylogenetic sense, is the condition of a shared attribute due to its presence in a common ancestor. This concept of homology and character is sufficient in a broad sense but does not differentiate between *trait homologs* and *character homologs*. Character homologs are those homologies that describe speciation events, and characters by definition, to be of cladistic value, must describe cladogenetic-speciation events. Phenetic definitions of character, and within-population traits, do not necessarily contain information pertinent to cladogenesis.

Lineages and Traits

The term *lineage* is widely used, often as a synonym of *monophyletic group*. In contrast, *lineage* is also utilized to describe groups of individuals within sexual populations that share a trait or combination of traits but that are not monophyletic, such as "maternal lineages" based on organellar traits (Palmer 1986; Avise 1986; Avise et al. 1987). In a sexual population, every individual is a member of numerous genetic lineages that may or may not "intersect" other lineages. The relationships of these lineages are not strictly hierarchic, cannot be described cladistically, and cannot be described using the terms *monophyly, paraphyly,* and *polyphyly* (Hennig 1966; Nixon, in press 1991). We shall identify two types of intrapopulational lineages—*minimal lineages,* which include all individuals that bear a trait or particular combination of traits, and *maximal lineages,* which include all descendants of a particular individual within a population or populations. Mitochondrial or chloroplast genotypes within populations are equivalent to minimal lineages, and as such are not equivalent to monophyletic groups of individuals. Maximal lineages within populations usually are *not detectable* because all descendants in a sexually recombining population do not bear all traits of their ancestors. Thus the "extinction" of a maximal lineage is ambiguous within sexual populations; the extinction of a minimal lineage is more important in terms of its effect on the distribution of traits. The death of the last individual with a particular trait is not necessarily the death of the last individual of a maximal lineage beginning with the original bearer of that trait. This trait extinction, however, has a greater implication for cladistics (and evolution) than does extinction of a maximal lineage, because the extinction of a trait removes it from the groundplan of all descendant species. A useful concept at this time is to group these two types of extinction—maximal lineage extinction and trait (or minimal lineage) extinction—together as microextinction.

Lineage Extinction and Speciation

Lineage extinction within sexual populations can be related directly to the formation of new phylogenetic species. Extinctions of minimal (trait) lineages within sexual populations can be viewed as a process of hierarchy formation. Within a sexual population, traits delimit incompletely intersecting sets of

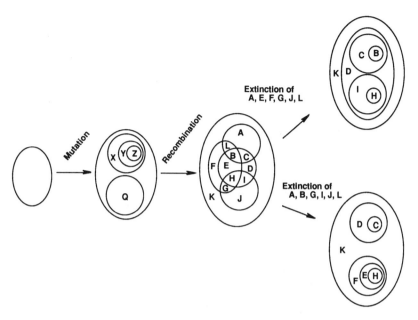

Fig. 4.4. Set representation of the effects of recombination and lineage extinction on sexual populations. Recombination disrupts hierarchic pattern, whereas removal (extinction) of particular lineages may result in reforming of hierarchic patterns.

elements (individuals) that are not related strictly in a nested hierarchy. By death of elements (individuals) that are members of set intersections, strictly hierarchic, nonhomoplasious patterns can emerge from these nonhierarchic patterns (fig. 4.4). Speciation, which is the local extinction of an alternative trait, can be viewed as a single step in the formation of a single character hierarchy across a set of elements. Repeated speciation events, or extinctions of traits, produce hierarchies that can be interpreted as phylogenies. The pattern of trait extinction determines the possible phylogenies that can arise from sexual populations with repeated speciation events. Overlain on this pattern are the continuing processes of genotypic origin, including things such as chromosomal rearrangement and deletion, and recombination, so that the nonhierarchic relations of elements (individuals) and distributions of traits within populations are dynamic.

It is important to note that high levels of lineage extinction may or may not be correlated with selection. The independence of "sorting" from selection has been pointed out by Vrba and Gould (1986), among others. "Random walks" are sufficient to explain speciation as a pattern, without the necessity of selection. Thus no particular model of lineage extinction (e.g., random or directed) is necessary to explain speciation and patterns of species diversity.

Clade Extinction

Extinction at higher levels (at and above that of species) is generally considered to be marked by the death of the last individual of a taxonomic group. Clearly, when taxa are not monophyletic, extinction may be an arbitrary notion, depending on how one defines the group in question. The popular notion of dinosaur "extinction" overlooks the evidence that modern birds and crocodiles belong to the same monophyletic group, and therefore should also be considered "dinosaurs" (Gauthier et al. 1988). Extinction above the species level takes on a more precise meaning when it is defined in terms of monophyletic lineages, or groups of species descended from a common ancestor. Other concepts of extinction may have utility from ecological or adaptational perspectives but lack rigor from a cladistic perspective.

Paraphyletic groups are clearly arbitrary entities that do not merit recognition in taxonomic systems. However, extinction also can be viewed on a character-by-character basis. When viewed in such a manner, the extinction of character state lineages that do not correspond to monophyletic groups may mark significant events. Plesiomorphic character states represent unmodified forms of particular characters, and the distribution of such character states may not correspond to monophyletic groups of species. The loss of plesiomorphic character states through extinction is irreversible and significant. Thus the extinction of the last "gymnosperm" would be significant in the sense that no new modifications of the gymnosperm strobilus could occur, except as modifications of the already modified angiosperm reproductive structures. The significance of extinction for evolution is related not only to the monophyletic nature of a group but also to the loss of unique character combinations.

Phylogenetic extinction in a general sense is the death of the last individual of a clade, thus removing a unique element from the universal set of species/ clades. We shall refer to such extinctions as clade or species extinctions. Because phylogenetic species (as discussed above) are marked by unique combinations of character states, species extinction also implies the loss (in living individuals) of these unique combinations of character states, and in many cases the complete loss of particular character states. The implications of such losses of character state combinations are numerous, but the most obvious result is the loss of a particular ancestral groundplan or matrix from which new species might arise. This is why when we speak of the biodiversity crisis (Wilson 1985, 1988), species extinction is only a part of the picture. Clade extinction, as is discussed in chapter 8, reduces diversity in terms of cladistic quality, not just species diversity. This situation is similar to the decimation concept of Gould (1989:49). Gould defined decimation as "reduction in the number of anatomical designs for life, not number of species." Is the loss of 20 orchid species as disturbing as the loss of each of the last species of 20 families of plants, animals, and/or fungi? Such questions are extremely important when assessing conservation priorities for the next few years, especially when public attention seems to focus most easily on vertebrates and smaller clades like the ones to which humans belong.

Clade extinction alters the subsequent possibilities for evolutionary change, just as trait extinction does within populations. The major difference is that clade extinction occurs in a clonal (phylogenetic) system, whereas lineage extinction within sexual populations is within a reticulate (tokogenetic) system. Thus clade extinction can have no direct effect on the evolutionary "potential" of related clades, in the sense of removing possible sources of traits (excluding hybridization). Trait extinction, however, changes possible evolutionary potentials for related lineages that do not bear the trait. Such is the nature of tokogenetic, or nonhierarchic, relationships. Mutation acts similarly within biparental tokogenetic systems, providing possible sources of traits for lineages that are independent at the time of the mutation.

The effect of clade extinction on cladistic analysis is analogous to the effect of lineage extinction on the pattern of character distribution in sexual populations. When the pattern of character distribution among a group of species is homoplasious, in a strict sense this pattern is not hierarchic. Cladistic analysis resolves a fully hierarchic pattern from a nonhierarchic pattern by reduction of homoplasy. If homoplasy occurs in the data, removal of elements (or extinction) can affect the pattern resolved through cladistic analysis, possibly changing the pattern found in the most parsimonious cladograms relative to the most parsimonious cladograms for the complete data set. Such changes could affect the inferred relationship of species not involved in the extinctions. Broadly interpreted, this means simply that adding or deleting taxa may change the results of a cladistic analysis if homoplasy occurs in the data. This obvious effect of the application of parsimony has been observed by numerous workers. Thus the discovery of a fossil may influence the outcome of a cladistic analysis (Donoghue et al. 1989). If character distributions are not homoplasious, extinctions will have no effect on the most parsimonious solution, and the inclusion of fossils will not change the pattern of the (single) most parsimonious cladogram of surviving species.

Fossils and Cladistic Analysis

Patterson has convincingly argued that properly analyzing characters is more important than appealing to the sequence of appearance of taxa in the fossil record, and concluded that (1981:220), "What remains is the unity of the comparative method, in which paleontology can hold its own by acknowledging its debt to neontology, and by repaying that debt in contributing what it alone can: age of groups, paleobiogeographic data, and new character combinations that can reverse decisions on homology and polarity, so testing, and perhaps on rare occasions overthrowing, theories of relationship."

Under some circumstances, fossils that lack parts and therefore lack character data may disrupt parsimony analyses, even if the character distributions for the preserved (known) characters are completely congruent with the most parsimonious trees found for only extant taxa. Thus a fossil with character states that partially match several nodes of the cladogram, if included in a parsimony analysis, will result in a multiplication of the number of parsimonious trees, and the fossil will appear as a "wildcard" taxon that is attached at different nodes to the same basic cladogram. Calculating strict (Nelson) consensus trees of such sets of cladograms will result in collapsed unresolved nodes (fig. 4.5) and the original structure of the cladogram will be lost. In the case of some fossils with missing data, it is therefore better to exclude the fossil from the analysis and then determine the range of nodes (groups) to which the fossil can be attached without affecting the length of the cladogram (Crepet and Nixon 1989a,b).

An important aspect of the use of fossils in cladistic analysis lies in the possibility of improving character hypotheses. Extinction, by removing "transitional" character states, can leave wide gaps in character data so that it is difficult to establish reasonable hypotheses of "character state adjacency." With the addition of fossil data, character state adjacencies and transformations may be more easily hypothesized. Such has been the case with fossil pollen of Fagaceae (beech family) in which fossils from the Oligocene of North America are intermediate in exine sculpture between extant members of modern subfamilies of Fagaceae (Crepet and Nixon 1989b). It is important to note that using fossils for determination of character state adjacency is not equivalent to using fossils for determination of character state polarity. Again, based on cladistic analyses, the oldest fossils may not be the most plesiomorphic, as is the case in the paleobotanical studies of Fagaceae. Fossil trigonobalanoids of Paleocene-Eocene age are more derived in some floral features than are fossil trigonobalanoids of Oligocene age. Thus chronology of appearance cannot be reliably used for determining character polarity (El-

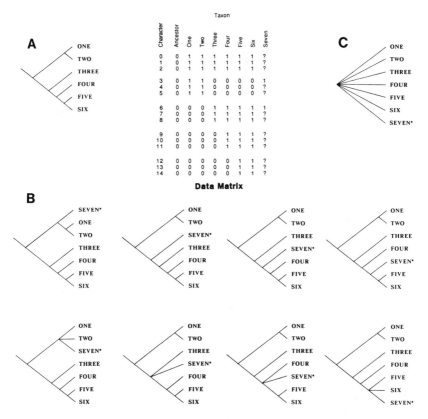

Fig. 4.5. The effect of "wildcard" taxa on cladistic analysis. Hypothetical data set, in which there is a single most parsimonious tree (A). (B) The addition of a hypothetical taxon with a large number of missing values results in eight equally parsimonious trees, all of which are identical except for the placement of the "wildcard" taxon ("seven*"). (C) The strict (Nelson) consensus tree of the eight trees that include the "wildcard" showing a complete deresolution of structure of the consensus tree.

dredge and Cracraft 1980; Patterson 1981; Nixon and Crepet 1989; Crepet and Nixon 1989a,b).

Anagenesis

Anagenesis is often discussed as a different kind of character change that occurs independently from cladogenesis (e.g., Hennig 1966). This "phyletic" change would be detected as correlated character changes along a branch of a cladogram, without independent clades emerging between successive character transformations. From the standpoint of Hennig (1966), anagenesis is character change that occurs within species, since his species concept equated the internodes of the phylogenetic tree with species. In contrast, with a Mayrian species concept, character change along a branch may be attributed to within-species anagenesis, or changes may be attributed to between-species change, depending on where species boundaries are drawn. Finally, from the standpoint of a phylogenetic species concept, character transformations always produce new species, so anagenesis cannot occur within species. We therefore prefer to describe anagenesis as successive character transformation (speciation) with extinction of ancestral species (fig. 4.6). Thus within-species anagenesis is clearly envisioned by Hennig, ambiguous with a Mayrian species concept, and interpreted as between-species character transformation with a phylogenetic species concept.

Although anagenesis (or its equivalent) is easily conceptualized in a cladistic framework, it seems impossible to distinguish between anagenesis as described earlier and the effects of extinction on our ability to detect other patterns of divergence. To eliminate the confounding effect of conflicting species concepts, anagenesis can be defined as successive character transfor-

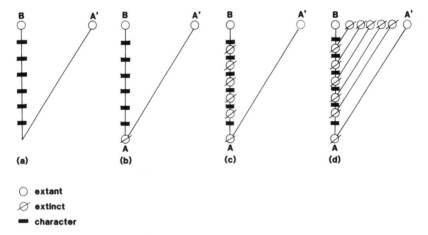

○ extant
⊘ extinct
■ character

Fig. 4.6. Diagramatic representation of "long branches" on cladograms, and possible interpretations of origins. (a) Observed pattern on cladogram. (b) Simultaneous transformation among six characters. (c) Successive speciation events involving character transformation, followed by extinction of "transitional" species. (d) Successive speciation, and cladogenesis, followed by extinction of "transitional" species as well as "side branches" derived from them.

mations that do not produce multiple coexisting species. This definition is compatible with the Hennigian, Mayrian, and phylogenetic species concepts. One important aspect of this definition is that it emphasizes the lack of branching inherent in an anagenetic pattern of change. Such a historical pattern of character change would result in correlated character transformations along a branch of a cladogram. Unfortunately, such patterns of correlated characters on cladograms could occur even if multiple species and clades were produced, followed by extinction of side branches (see fig. 4.6d).

Simultaneous character transformation among two or more characters is another possible source of correlated character transformations on a cladogram (e.g., fig. 4.6b). Simultaneous character transformation occurs under only one condition: when a single individual bears the plesiomorphic state for all the correlated characters, and the death of that individual marks the transformation of all the characters, resulting in speciation. Under this model, extinction of the ancestral species is not necessary for anagenesis to occur. Again, from the standpoint of the phylogenetic species concept, this form of anagenesis is not interpreted as within-species change, although it might be interpreted as such with a Mayrian or Hennigian species concept.

Because of these problems, anagenesis must be viewed as a particular pattern of character transformation that corresponds to more than one process that cannot be detected with cladistic analysis. This does not deny the possibility that anagenesis, as broadly defined previously, has occurred. What may appear to be anagenetic change along a branch of a cladogram may be an artifact of extinction that is not assignable to a source. We can see no way of distinguishing between the alternative explanations other than with particular assumptions or models of speciation, character transformation, and extinction.

Assuming no simultaneous transformation of characters and complete knowledge of characters, the maximum number of extinct (phylogenetic) species within a clade can be taken to approximate the number of character transformations (see fig. 4.6c). This is, of course, necessarily imprecise. Without knowledge of genetically or developmentally correlated transformations, the number of such "species" might be overestimated. Because every data set almost certainly includes only some fraction of the actual number of characters, such a number is almost always an underestimate. And unless unit characters can be delimited, these numbers are rarely directly comparable across clades. Nonetheless, this does provide a crude, tentative estimate of the relative past diversity of clades. As an example, the 21 autapomorphies of the lymexylid beetle genus *Atractocerus* suggest a considerable level of past diversity and extinction (Wheeler 1986).

Anagenesis and cladogenesis can be viewed also in terms of lineages. If one wishes to define the origins of within-species lineages arbitrarily, then the death of every individual is the extinction of at least one lineage. Such definitions of extinction are themselves arbitrary and of doubtful use to systematics. Extinction of traits, however, may mark important events and irreversibly change the course of tokogenetic and phylogenetic history. The local extinction of a trait excludes that trait from subsequent evolutionary events associated with that "local" part of living things. All relevant extinction involves the loss of character-trait information. When extinction of a plesio-morphic trait is associated with fixation of an apomorphic trait, detectable cladogenesis, or speciation, has occurred. If the plesiomorphic trait is com-pletely eliminated throughout all populations, speciation might be considered to be "sympatric" or "anagenetic" (fig. 4.7a). If the trait is eliminated only in a particular population, speciation might be considered to be "allopatric" or "cladogenetic" (fig. 4.7b). However, these distinctions cannot be made inter-nally on a cladogram, as "anagenesis" and "cladogenesis" with subsequent extinction of plesiomorphic populations result in identical patterns (see fig. 4.6). When dealing with terminal taxa (phylogenetic lineages), one can say that a "metaspecies"-species pair that differs by a particular apomorphy is not

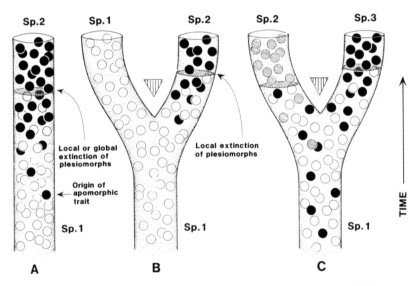

Fig. 4.7. Graphic representation of speciation from the perspective of a phylogenetic species concept. (a) "Anagenetic" speciation, with simultaneous extinction of ancestral species. (b) and (c) "Cladogenetic" speciation, resulting in one (b) or two new species (c), with the extinction of the ancestral species.

the result of "anagenetic" speciation, since the "ancestral" species is still extant (see fig. 4.7b). One cannot say that sister species marked by two different apomorphies are "anagenetic," since for two species to develop from a common ancestor, the ancestral matrix cannot become extinct with the first speciation event.

Implications of Extinction for Evolutionary Models

Set theory provides a framework for evaluating the implications of extinction in relation to cladistic analysis. This allows the identification of significant historical events that have resulted in the observed pattern of species diversity. These events are the extinction of plesiomorphic attributes simultaneous with the fixation of apomorphic attributes, resulting in character transformation, speciation, and cladogenesis. The phylogenetic species concept is not only desirable but necessary for understanding these patterns. This unified view of phyletic evolution provides a basis to evaluate models of evolutionary change. Of particular interest are models that deal with rates of evolution, anagenesis, and stasis.

Cracraft (1985) discussed extinction from the standpoint of biological diversity and proposed that both speciation and extinction rates are independent of standing diversity. He further postulated that "speciation rate is controlled primarily by large-scale changes in lithospheric (geomorphologic) complexity." These ideas are fully compatible with the model of speciation that we have proposed, in that changes in geomorphologic complexity will result in populations being fractured, increasing rates of lineage extinction within populations, and, ultimately, increased speciation.

Saltational models of evolution are based on negative evidence—the lack of transitional forms. Anagenesis, or the occurrence of a long branch in a cladogram, is a phenomenon that can be the result of two distinct patterns of historical events—lack of diversification of clades, or extinction of clades. Saltational models are hypotheses that "transitional" species never existed. Fossil data can falsify saltational hypotheses by providing evidence of gradualism, or "single-character" transformations, through the occurrence of intermediate transitional forms.

The classic distinction of rates of speciation as comparatively horotelic, bradytelic, or tachytelic (Simpson 1944) are potentially tractable under the model we describe, where rates of character transformation (fixation) are more or less equal to rates of speciation. Such rates, however, are only meaningful when "comparable" clades are used. Comparability must be based on a standardization along a time scale. That is, groups to be compared must originate at the same time. This, of course, is fully unambiguous only for sister taxa (see essay 3, for further discussion and references).

It has been noted that in some clades "high extinction rates are accompanied by high origination rates at most times" (Flessa et al. 1986). Such patterns are to be expected when lineage extinction is high, because speciation and species extinction are simply alternative results of lineage extinction. Whether the pattern of lineage extinction is fully stochastic or directed in some manner, high rates of lineage extinction will eliminate some species, transform some species into new apomorphic forms, and "break up" some species into component populations (new species) fixed for different character states. Arguments that habitats are "opened up" following mass extinction are unnecessary for explanations of high "rediversification" rates. If a hypothetical lineage extinction curve is visualized, any gradual fluctuation in lineage extinction will result in high (relative) rates of diversification preceding and following the highest rates of species and population extinction. Such an explanation is independent of particular models of evolution, based only on the observable effect of lineage extinction on sexual populations. In asexual organisms, which do not undergo recombination, high rates of lineage extinction will not be associated with high rates of rediversification. This suggests that sexually recombining systems are more resilient to the effects of extinction than are asexual populations, a conclusion that has been proposed on numerous other grounds.

The punctuated equilibrium model of Eldredge (see Eldredge and Gould 1972) describes a pattern of long periods of "stasis" of species followed by periods of rapid anagenetic change and diversification. This model is clearly related to the observation of high rediversification rates (Flessa et al. 1986), as noted earlier. Although we may disagree with characterizations of anagenetic change, the punctuated equilibrium model is neither supported nor refuted by our model of extinction and hierarchy. Factors affecting levels of lineage extinction would also affect levels of cladogenesis (and observed "anagenesis" in cladograms). Low rates of lineage extinction also would mean low levels of cladogenesis; thus species would persist for long periods, and fewer new species would arise. High rates of lineage extinction would produce high rates of species and clade extinction and high rates of speciation as well; these would be the "punctuation" points. During periods of high lineage extinction, there would be many more speciation events, but species would be short-lived. It seems to be difficult to separate these factors—in other words, it is hard to envision high rates of speciation and low rates of species extinction, given our model. This interdependence is not an equilibrium, but instead two independent consequences of the same factor. Although our model predicts an association between extinction and speciation, it does not shed any light on the causes of cyclic fluctuations in rates as proposed in the punctuated equilibrium model. Explanations for differing rates of lineage extinction must rely on models of processes that are independent of cladistic theory, and in general are not testable with cladistic methods.

ACKNOWLEDGMENTS

We thank Jerrold Davis, Melissa Luckow, and the systematics graduate students of the Bailey Hortorium and Department of Entomology at Cornell University, for numerous discussions of the topics presented in this chapter. We thank Michael Novocek for comments on the manuscript.

REFERENCES

Avise, J. C. 1986. Mitochondrial DNA and the evolutionary genetics of higher animals. *Phil. Trans. R. Soc. Lond. B* 312:325–342.

Avise, J. C., J. Arnold, R. M. Ball, E. Bermingham, T. Lamb, J. E. Neigel, C. A. Reeb, and N. C. Saunders. 1987. Intraspecific phylogeography: The mitochondrial DNA bridge between population genetics and systematics. *Ann. Rev. Ecol. Syst.* 18:489–522.

Brundin, L. 1972. Evolution, causal biology, and classification. *Zool. Scripta.* 1:107–120.

Cracraft, J. 1983. Species concepts and speciation analysis. In R. F. Johnston, ed., *Current Ornithology,* 1:159–187. New York: Plenum.

Cracraft, J. 1985. Biological diversification and its causes. *Ann. Missouri Bot. Gard.* 72:794–822.

Crepet, W. L. and K. C. Nixon. 1989a. Earliest megafossil evidence of Fagaceae: phylogenetic and biogeographic implications. *Amer. J. Bot.* 76:842–855.

Crepet, W. L. and K. C. Nixon. 1989b. Extinct transitional Fagaceae from the Oligocene and their phylogenetic implications. *Amer. J. Bot.* 76:1493–1505.

Darwin, C. 1859. *On the Origin of Species.* London: J. Murray.

Diamond, J. M. 1984. "Normal" extinctions of isolated populations. In M. H. Nitecki, ed., *Extinctions,* pp. 191–246. Chicago: University of Chicago Press.

Donoghue, M. J., J. Doyle, J. Gauthier, A. Kluge, and T. Rowe. 1989. The importance of fossils in phylogeny reconstruction. *Ann. Rev. Ecol. Syst.* 20:431–460.

Eldredge, N. and J. Cracraft. 1980. *Phylogenetic Patterns and the Evolutionary Process: Method and Theory in Comparative Biology.* New York: Columbia University Press.

Eldredge, N. and S. J. Gould. 1972. Punctuated equilibria: an alternative to phyletic gradualism. In T. J. M. Schopf, ed., *Models in Paleobiology,* pp. 82–115. San Francisco: Freeman, Cooper.

Farris, J. S. 1979. The information content of the phylogenetic system. *Syst. Zool.* 28:483–519.

Farris, J. S. 1982. Outgroups and parsimony. *Syst. Zool.* 31:328–334.

Farris, J. S. 1983. The logical basis of phylogenetic analysis. In N. I. Platnick and V. A. Funk, eds., *Advances iin Cladistics,* 2:7–36. New York: Columbia University Press.

Flessa, K. W., H. K. Erben, A. Hallam, K. J. Hsu, H. M. Hussner, D. Jablonski, D. M. Raup, J. J. Sepkoski, Jr., M. E. Soule, W. Sousa, W. Stinnesbeck, and G. J. Vermeij. 1986. Causes and consequences of extinction. In D. M. Raup and D. Jablonski, eds., *Patterns and Processes in the History of Life,* pp. 235–257. Berlin: Springer-Verlag.

Gauthier, J., Kluge, A. G., and T. Rowe. 1988. Amniote phylogeny and the importance of fossils. *Cladistics* 4:105–209.

Gould, S. J. 1989. *Wonderful Life.* New York: Norton.

Hennig, W. 1966. *Phylogenetic Systematics.* Urbana: University of Illinois Press.

Kluge, A. G. 1989. Metacladistics. *Cladistics* 5:291–294.

Mayr, E. 1942. *Systematics and the Origin of Species.* New York: Columbia University Press.

Mayr, E. 1963. *Animal Species and Evolution.* Cambridge: Harvard University Press.

Nelson, G. 1978. Ontogeny, phylogeny, paleontology, and the biogenetic law. *Syst. Zool.* 27:324–345.

Nelson, G. 1989a. Species and taxa: Systematics and evolution. In D. Otte and J. A. Endler, eds., *Speciation and Its Consequences,* pp. 60–81. Sunderlin, Mass.: Sinauer.

Nelson, G. 1989b. Cladistics and evolutionary models. *Cladistics* 5:275–289.

Nelson, G. and N. I. Platnick. 1981. *Systematics and Biogeography: Cladistics and Vicariance.* New York: Columbia University Press.

Nixon, K. C. in press. Monophyly, paraphyly, and tokogeny: an assessment of hierarchic and non-hierarchic patterns of descent. *Cladistics.*

Nixon, K. C. and W. L. Crepet. 1989. *Trigonobalanus* (Fagaceae): taxonomic status and phylogenetic relationships. *Amer. J. Bot.* 76:828–841.

Nixon, K. C. and Q. D. Wheeler. 1990. An amplification of the phylogenetic species concept. *Cladistics* 6:211–223.

Palmer, J. D. 1986. Chloroplast DNA and phylogenetic relationships. In S. K. Dutta, ed., *DNA Systematics, Vol. II: Plants.* Boca Raton: CRC Press.

Paterson, H. E. H. 1985. The recognition concept of species. In E. Vrba, ed., *Species and Speciation.* Transvaal Museum Monograph 4:21–29.

Paterson, H. E. H. and M. Macnamara. 1984. The recognition concept of species. *South African J. Sci.* 80:312–318.

Patterson, C. 1981. Significance of fossils in determining evolutionary relationships. *Ann. Rev. Ecol. Syst.* 12:195–223.

Patterson, C. 1982. Morphological characters and homology. In K. A. Joysey, and A. E. Friday, eds., *Problems of Phylogenetic Construction.* London: Academic Press.

Pimentel, R. A. and R. Riggins. 1987. The nature of cladistic data. *Cladistics* 3:201–209.

Platnick, N. I. 1979. Philosophy and the transformation of cladistics. *Syst. Zool.* 28:537–546.

Queiroz, de, K. and M. J. Donoghue. 1988. Phylogenetic systematics and the species problem. *Cladistics* 4:317–338.

Queiroz, de, K. and M. J. Donoghue. 1990. Phylogenetic systematics or Nelson's version of cladistics? *Cladistics* 6:61–75.

Quine, W. 1961. *Mathematical Logic.* Rev. ed. Cambridge: Harvard University Press.

Raup, D. M. 1984. Death of species. In M. H. Nitecki, ed., *Extinctions,* pp. 1–19. Chicago: University of Chicago Press.

Ridley, M. 1989. The cladistic solution to the species problem. *Biol. Philos.* 4:1–16.

Rosen, D. E. 1978. Vicariant patterns and historical explanation in biogeography. *Syst. Zool.* 27:159–188.

Rosen, D. E. 1979. Fishes from the uplands and intermontane basins of Guatemala: revisonary studies and comparative geography. *Bull. Amer. Mus. Nat. Hist.* 162:267–376.

Simpson, G. G. 1944. *Tempo and Mode in Evolution.* New York: Columbia University Press.

Vrba, E. S. and S. J. Gould. 1986. The hierarchical expansion of sorting and selection: Sorting and selection cannot be equated. *Paleobiology* 12:217–228.

Watrous, L. E. and Q. D. Wheeler. 1981. The out-group comparison method of character analysis. *Syst. Zool.* 30:1–11.

Wheeler, Q. D. 1981. The ins and outs of character analysis: A response to Crisci and Stuessy. *Syst. Bot.* 6:297–306.

Wheeler, Q. D. 1986. Revision of the genera of Lymexylidae (Coleoptera: Cucujiformia). *Bull. Amer. Mus. Nat. Hist.* 183:113–120.

Wheeler, Q. D. 1990a. Ontogeny and character phylogeny. *Cladistics* 6:225–268.

Wheeler, Q. D. 1990b. Morphology and ontogeny of postembryonic larval *Agathidium* and *Anistoma* (Coleoptera: Leiodidae). *Amer. Mus. Novitates,* 2986, 46 pp.

Wilson, E. O. 1985. The biological diversity crisis: a challenge to science. *Issues Sci. Technol.* Fall: 20–29.

Wilson, E. O., ed., 1988. *Biodiversity.* Washington: National Academy of Science Press.

Woodger, J. H. 1937. *The Axiomatic Method in Biology.* London: Cambridge University Press.

Wright, S. 1982. Character change, speciation, and higher taxa. *Evolution* 36:427–443.

5 : Trilobite Phylogeny and the Cambrian-Ordovician "Event": Cladistic Reappraisal

Gregory D. Edgecombe

Abstract. Cladistic revision of supergeneric taxa of Trilobita has revealed that taxonomic turnover between Cambrian and Ordovician systems is less than was implied by traditional gradistic classifications. Taking the Late Cambrian "Ptychaspid Biomere" as an exemplar of biomere diversification, taxonomic artifacts (nonmonophyletic groups) have contributed to spurious extinction patterns: cladistic relationships predict ages of origination (sensu Hennig) that may predate biomere boundaries. Calibrating a cladogram for the suborder Asaphina with stratigraphic first appearances, ghost lineages (range extensions inferred from earlier appearance of a sister group) are a substantial component of Cambrian taxon duration. Characters from protaspid (larval) morphology afford resolution of the enigmatic relationships of several highly autapomorphic trilobite orders that diversified in the Early Ordovician: the order Phacopida provides a case study. A high degree of taxonomic congruence is afforded between cladograms based on larval and adult characters for phacopide families.

Much has been said in recent years about gross patterns of diversification of taxa through time. Because of their long and comparatively well-preserved fossil record, marine shelly faunas have been of particular interest. It has been suggested that changing taxonomic composition in the Phanerozoic allows three "evolutionary faunas" to be identified (Sepkoski 1979, 1981), the Cambrian, Paleozoic, and Modern faunas. The relationship between the first two of these "phases of diversification" is of special interest to trilobite taxonomists. Sepkoski's data base for the "Cambrian fauna" is composed in large part of Trilobita, an arthropod clade whose species diversity peaked in the Late Cambrian. The Cambrian-Ordovician boundary is cited as a peak in extinctions (e.g., some 70% of marine genera allegedly suffered this fate; Sepkoski 1986).

The question of turnover in trilobite lineages relates to similar issues in other marine groups. In a consideration of cladistic relationships of echinoderms across this systemic boundary, Smith (1988) suggested that taxonomic artifacts exaggerate apparent patterns of Cambrian extinction and Ordovician origination. He discovered that many Cambrian echinoderm taxa have been classified in paraphyletic groups, parts of which have Ordovician sister taxa. In many cases, stratigraphic separation between sister groups demands substantial range extensions beyond those predicted by the stratigraphic first occurrence of clades. The significance of sister-group relationships for inferring age of origination of taxa is discussed in detail by Norell in chapter 3.

To distinguish this method of inference from Flessa and Jablonski's (1983) "Lazarus effect" (which simply corrects for gaps in stratigraphic occurrence), I follow Hennig's framework for age of taxonomic origination. "Age of origin" (T1) of Hennig (1965, 1966) is (minimally) estimated by the known first occurrence of a sister taxon. Hennig's (1965) "age of division" (T3) (called "age of differentiation" [T2] by Hennig in 1966) referred to the age at which the latest stem species of all Recent species of the group split. The emphasis Hennig placed on age of differentiation reflects his view that fossils play a secondary role in the phylogenetic system. The search for the most inclusive group of Recent species foreshadows Hennig's (1969) "*gruppe" and "stammegruppe" conventions (crown and stem groups of Jefferies 1979). "Age of division" (differentiation) is irrelevant to completely extinct clades such as Trilobita. Moreover, the conceptualization of branching diagrams in terms of stem species and their division (e.g., Ax 1987) is appropriate to analysis at the level of phylogenetic trees, but not necessarily at the (more general) analytical level of cladograms (Nelson and Platnick 1981). Cladistics affects stratigraphic analysis of survivorship by distinguishing T1 from the observed first appearance of a clade; the intervening taxonomic history has been called a "ghost lineage" (Norell, chapter 3 of this volume). Smith (1988) has taken evidence from Lazarus taxa and revision of paraphyla to imply that enchinoderms do not support the two-phase (Cambrian and Paleozoic) pattern of diversification predicted by traditional taxonomies. His methods do not consider the additional correction for ghost lineages.

The question, then, is "What temporal patterns of diversification are evident for trilobites?" The query is important because trilobites have a Cambrian record that is more complete than that of echinoderms. There is no dispute that the Cambrian-Ordovician boundary is the equivalent of (or at least has been interpreted as the equivalent of) a taxonomic boundary for many groups. Joachim Barrande in 1846 noted that changes in trilobite morphology are a basic criterion for distinguishing the *faune primordiale* and *faune seconde* (a distinction that encouraged Lapworth's later recognition of a three-system division of the Lower Paleozoic). As evidence for the endurance of this notion, consider H. B. Whittington's (1954) reference to a great gap between taxa on opposite sides of the Cambrian-Ordovician boundary (cf. Foote 1989). Whittington's taxonomic summary (1954: Fig. 1) recognizes extinction of many components of a nonmonophyletic Cambrian ptychoparioid group, depicted as ancestral to a polyphyletic grouping of post-Cambrian taxa. Other diverse post-Cambrian clades had an undetermined

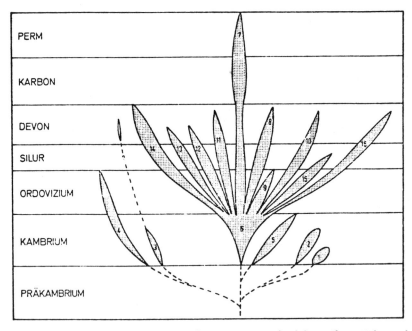

Fig. 5.1. Traditional "star phylogeny" of major groups of trilobites (from Hahn and Hahn 1975). "Cryptogenetic" post-Cambrian orders 7–16 are depicted as radiating from Cambrian Ptychopariida (taxon 6) with unspecified interrelationships. Groups considered here are Proetina (7), Asaphida (9), Lichida (10), Odontopleurida (11), Calymenida (12), Cheirurida (13), and Phacopida (14).

relationship to Cambrian taxa. This pattern of distinctive new clades originating in the Early Ordovician was dubbed "cryptogenesis" by Stubblefield (1959). Its essence is summarized in Hahn and Hahn's (1975) phylogenetic tree for Trilobita (fig. 5.1), with 10 ordinal groups originating from separate ptychopariide ancestries across the Cambrian-Ordovician boundary.

In short, post-Cambrian trilobite orders have been of largely unknown relationship to each other and to Cambrian taxa. Their depiction as descendants of an ancestral Cambrian order Ptychopariida can hardly be regarded as informative. The lack of hierarchic resolution can result as much from ignorance of relationships as from an apparent explosive evolutionary burst in the Early Ordovician following Cambrian extinction (cf. Edgecombe 1991, on the "Cambrian explosion"). Cladistic relationships can be proposed between several of the orders traditionally conceived as post-Cambrian groups, with the argument that certain Cambrian taxa are included within these clades. The effect of revised phylogenies is to extend known ages of origination of groups, which, in part, dampens the "Ordovician radiation."

Cambrian Trilobite Extinction and the Biomere Model: Cladistics and Survivorship Patterns

A first problem to consider is the extent of the latest Cambrian trilobite extinction. Here one may acknowledge potential artifacts based on uncertainties in correlating the base of the Ordovician System on a global scale (Henningsmoen 1973; Norford 1988). A stratotype has not been selected and, even when it is, it will remain difficult to correlate between different paleogeographic provinces (e.g., the British Tremadoc Series and the Ibexian series on the North American platform). It is, however, widely accepted that the Cambrian-Ordovician boundary in North America closely corresponds to the upper boundary of a biostratigraphic unit called a biomere. A. R. Palmer, who first outlined the biomere concept, attempted to explain an apparent repeating pattern of diversification and mass extinction observed in late Cambrian trilobites (Palmer 1965a). Biomeres were defined as packages of strata bounded by extinction of the "dominant elements of a single phylum." These intervals are not necessarily related to physical breaks in the sedimentary record and are defined by possibly diachronous boundaries. The claim of diachroneity, distinguishing the biomere as a biostratigraphic unit, was questioned by Henderson (1976) and later dismissed by Palmer (1984:605). Palmer's model (based on what was named the Pterocephaliid Biomere: Palmer 1965b) invokes a deep-water oceanic stock invading the shelf, evolving in situ, and suffering mass extinction.

Other Cambrian trilobite workers have found the biomere concept to be useful, and additional biomeres have been documented. Stitt (1971, 1975, 1977) cited evidence for a younger Ptychaspid Biomere and an older Marjumiid Biomere (Conaspid and Crepicephalid Biomeres of Palmer 1965b), each named after a particular trilobite family whose abundance was thought to be characteristic of the unit. The Lower and Middle Cambrian have been informally "biomerized" (Palmer 1981), as the "Olenellid" and "Corynexochid" Biomeres; the Ibexian (very latest Cambrian and Early Ordovician) has been termed a "Symphysurinid Biomere" (Stitt 1983; Hystricurid Biomere of Stitt 1977). Stitt's interpretation of the biomere pattern (1975: fig. 1) elaborates Palmer's original model; he outlines a four-stage scenario of adaptive radiation.

The biomere model of radiation (fig. 5.2) is readily subject to cladistic testing. If the phylogenetic implications of the biomere are correct, we should find that trilobites of a biomere comprise a monophyletic group and that none is more closely related to taxa in earlier or in younger strata. A survey

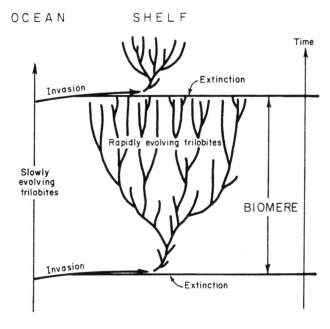

Fig. 5.2. Phylogenetic model of a biomere as an in situ adaptive radiation (from Stitt 1977). A prediction of the model is that the trilobite fauna of the biomere is a monophyletic assemblage.

of cladistic relationships for trilobites of the Ptychaspid Biomere (Sunwaptan Stage of Ludvigsen and Westrop 1985) suggests that this is not the case. Certain families known only from this interval are nested within monophyletic groups with Middle Cambrian representatives or have sister groups ranging into the Ordovician. The discovery that biomeres are not monophyletic radiations renders dubious the calculations of evolutionary rates for Late Cambrian trilobites (Ludvigsen 1982; Westrop 1990). For example, Stanley (1979) computed Palmer's trilobites as providing one of the only examples of an invertebrate group showing rates of speciation as rapid as those of mammals. These dramatic rates derive from assuming a single (or, at most, a few) ancestral species at the biomere's base giving rise to all polymerid trilobites in the Pterocephaliid Biomere in the Great Basin of western North America. Such calculations of diversification rate are greatly affected when one must account for Ptychaspid Biomere taxa (e.g., "Saukiidae") being more closely related to Swedish or Chinese Middle Cambrian anomocarids than to co-occurring taxa such as plethopeltids (ghost lineages can be used to recalibrate minimum age of divergence). Likewise, Hardy (1985) sought an explanation of why Ptychopariida of the Ptychaspid Biomere fail to conform to predictions for adaptive radiations. An explanation may simply lie in the argument that Ptychopariida is paraphyletic (Eldredge 1977). Nonphylogenetic groupings are not required to behave as evolutionary theories predict.

The preceding statements will be reviewed in light of some evidence. First, are biomere boundaries empirical phenomena? If so, terminations of species ranges should be nonrandomly concentrated at the alleged extinction level. This appears dubious for the Ordovician Symphysurinid Biomere; only one species has a last appearance at the claimed extinction level, the uppermost boundary of the *Paraplethopeltis* zone (Stitt 1983: pl. 7). Stitt's data for the Ptychaspid Biomere (1971: fig. 2), from carbonate platform sites in southern Oklahoma, are at least consistent with a "mass extinction" interpretation of biomere boundaries. However, sites positioned closer to the continental margin, such as Ludvigsen's (1982) data for sections in northwestern Canada, show a more ambiguous pattern of staggered and overlapping ranges (see Briggs et al. 1988: fig. 9.4). Palmer (1984) reported nearly knife-sharp resolution of the Marjumiid and Pterocephaliid Biomere boundaries, but Westrop and Ludvigsen (1987) observe the "critical period" at the top of the Ptychaspid Biomere to range through as much as 26 m of strata. Evidently, there is biogeographic control of resolution of biomere boundaries. On the one hand, taphonomic effects might bias the apparent sharp turnover in shallow settings. Alternatively, since different species are represented in these sections, the difference might be ecological (e.g., platform taxa might be more vulnerable to environmental perturbation). Existing data suggest that the abruptness of turnover diminishes in successively younger biomeres (F. A. Sundberg personal communication).

Allowing that the Cambrian-Ordovician boundary closely corresponds to one of the aforementioned turnover "events" (the traditional Cambrian Trempealeauan Stage extends just above the top of the Ptychaspid Biomere), we might examine the taxonomic basis for the in situ adaptive radiation model. (I will avoid the popular issues concerning the cause of extinctions (see Westrop and Ludvigsen 1987; Fortey 1989) and the likelihood that morphologies are iterated between biomeres (Sundberg 1989). As a case study, consider a proposed phylogeny of the nominate family of the Ptychaspid Biomere (fig. 5.3). Longacre (1970) revised Ptychaspididae as a clade restricted to the Ptychaspid Biomere, including four subfamilies. Lochman's (1956: fig. 3) phylogenetic tree for the taxa differs in recognizing a eurekiid–saukiid clade,

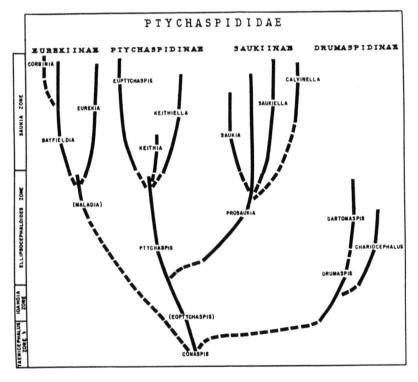

Fig. 5.3. Phylogeny of Ptychaspididae (after Longacre 1970), interpreted as a monophyletic group with origination and extinction in the "Ptychaspid Biomere."

with ptychaspidids a paraphyletic group (fig. 5.4B). Inclusion of the subfamily Drumaspidinae (Longacre 1970) is novel: it was justified without definitive character support, but with emphasis on spatiotemporal association. Based solely on morphology, Westrop (1986) argued that drumaspidines are most closely related to Elviniidae (Olenacea?), which biomere taxonomists conceived as restricted to the underlying Pterocephaliid Biomere (Palmer 1965b). A hypothesis of within-biomere monophyly explains this similarity as homoplastic and forces pseudoextinction of elviniids (excepting the genus *Irvingella*) at the top of a biomere.

Palmer (1965b:33) criticized solely morphologic classifications, such as that of the trilobite Treatise, which do not allow spatial and temporal "characters" to be considered in resolving relationships. This reasoning, however, invites circularity (biomeres defining monophyly) and allows no independent historical discoveries of distribution patterns (cf. Platnick 1985). The assertion that morphology alone cannot distinguish "morphologic similarities that may indicate real genetic relationships . . . from morphologic similarities resulting from convergence or happenstance" (Palmer 1965b:33) is opposed by a homology concept based on maximum congruence of derived topological similarities (Rieppel 1988).

Biomere phylogeneticists thus advocate that these stratigraphic units are closed extinction-bounded systems. A more moderate stance (Stitt 1983:16) allows that taxa may be, at most, minor elements in biomeres above or below that in which they are characteristic. Lochman's and Longacre's ptychaspidid phylogenies are presented as ancestor-descendant trees, rather than as cladograms (fig. 5.4A,B); monophyly of the group is violated, as supposedly unrelated clades are nested within the tree. As an example, consider the relationships of the Late Cambrian family Dikelocephalidae to the ptychaspidids. The superfamily Dikelocephalacea (fig. 5.4C) has been revised (Ludvigsen and Westrop 1983; Westrop 1986) as a monophyletic group including only two of the three or four clades depicted in the "biomere-defined" ptychaspidids (Saukiinae and Ptychaspidinae); one of them (Saukiinae) is most closely related to a family (Dikelocephalidae) considered unrelated by advocates of the biomere model. A stratigraphic argument was used to exclude Dikelocephalidae from a saukiid-ptychaspidid grouping (Lochman 1956:450). Saukiids share more derived characters with dikelocephalids than with ptychaspidids, but existing diagnoses of the group are gradistic. Accordingly, "Saukiidae" is an ill-defined paraphyletic group (Ludvigsen et al. 1989).

"Saukiid" extinction is thus reinterpreted as a pseudoextinction. Last occurrences of such paraphyla can only be considered as informative (Smith and Patterson 1988:145) if part of the paraphyletic group outranges its monophyletic sister group. In this instance, one "saukiid" clade (*Mictosau-*

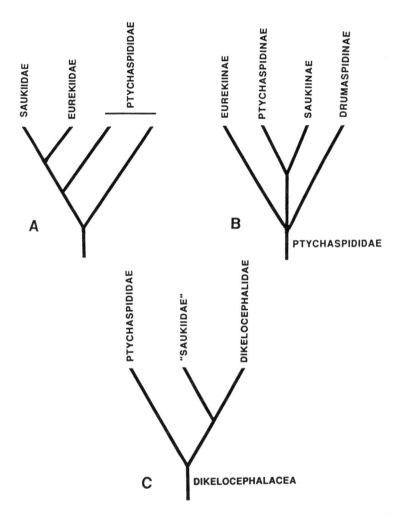

Fig. 5.4. (A) Cladistic relationships between Ptychaspididae, Eurekiidae, and Sauki-idae (after Lochman 1956: fig. 3); (B) Cladistic relationships between subfamilies of Ptychaspididae (after Longacre 1970: fig. 1); (C) Cladistic relationships of Dikelo-cephalacea (sensu Ludvigsen and Westrop 1983). Note that eurekiids (Remopleurida-cea) and drumaspidines (Elviniidae) are excluded; Dikelocephalidae is included as sister group to (paraphyletic) "Saukiidae."

kia—Shergold 1975) persists into the Early Ordovician (Tremadoc) in Mexico and Afghanistan. [This occurrence may, however, be within Westrop's (1989) "critical period" of extinction.] Regarding first occurrence, "saukiids" are known from Australia in the early Late Cambrian Mindyallan Stage (Öpik 1967). This stratigraphic appearance (equivalent to the upper part of the Marjumiid Biomere) can be used to calibrate a minimum age for divergence of Ptychaspididae from the "saukiid"–dikelocephalid clade. As such, Dikelocephalacea is not restricted by the lower (and possibly the upper) boundary of the Ptychaspid Biomere as predicted by an in situ radition model; major clades differentiated well before that biomere's base and predate *Conaspis*, often depicted as the ancestral taxon (see fig. 5.3). Nor, we shall see, is the sister group of the dikelocephalaceans restricted to the biomere. There is evidence to suggest that Dikelocephalacea is part of a more inclusive group of Trilobita, one with a known stratigraphic range from the Middle Cambrian to the Upper Silurian. By contrast, Eurekiidae, regarded as part of the ingroup for the ptychaspidid radiation by Longacre (1970), appears to have closest relations to taxa other than Dikelocephalacea; Ludvigsen et al. (1989) cite evidence for classifying this group as Remopleuridacea.

It has been suggested that a number of Cambrian-Ordovician families can be united based on a unique condition of ventral cephalic sutures (Fortey and Chatterton 1988). The primitive state for "ptychopariides" is presence of a rostral plate between paired connective sutures. Asaphina sensu Fortey and Chatterton (including our example, Dikelocephalacea) is distinguished by a ventral median suture (without an intervening rostral plate). Support for monophyly of this group is provided by congruence with a unique life history pattern. Asaphina has a distinctive planktonic larval (protaspid) morphology, with a dramatic metamorphosis into adultlike instars of the meraspid period (Speyer and Chatterton 1989). The plesiomorphic pattern for "Ptychopariida" is gradual (nonmetamorphic) transformation from protaspis larva into meraspis (the boundary between these stages in the trilobite life cycle is when the cephalon and transitory pygidium first become separate, articulated sclerites).

This conjecture of monophyly invites parsimony analysis of asaphine families, for which I draw on recent investigations by Fortey and Chatterton (1988). Slight changes have been made to their published data set (their table 3; see fig. 5.5); 40 exoskeletal characters are coded for 10 ingroup terminal taxa using "ptychopariide" outgroups. With multistate characters nonadditive (i.e., unordered transformations), the implicit enumeration (ie*) option of HENNIG 86 (Farris 1988) retrieves four shortest cladograms of 65% consistency. Figure 5.5 is based on a strict consensus of these cladograms. Observe that dikelocephalaceans are resolved as a sister group to Remopleuridacea, which ranges from the Late Cambrian to the end of the Ordovician. The

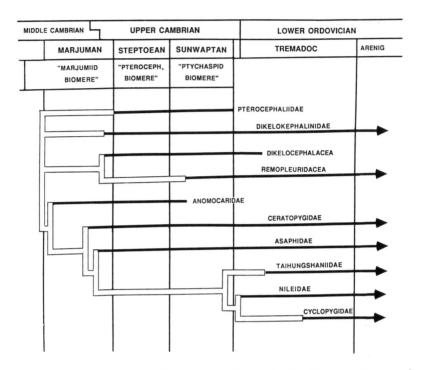

Fig. 5.5. Stratigraphic context of sister-group relations for Asaphina sensu Fortey and Chatterton (1988). Strict consensus of four minimum-length cladograms (CI = 0.65) generated from character data of Fortey and Chatterton (1988: Tables 2 and 3) with implicit enumeration option (ie★) of HENNIG 86 (Farris 1988). Multistate characters were coded as nonadditive. Characters 9 and 21 were deleted (character 9 is redundant on character 8; character 21 is not coded discretely). Solid vertical bars are observed stratigraphic ranges of terminal taxa; empty bars ("ghost lineages" sensu Norell, chapter 3 of this volume) are inferred range extensions based on known occurrence of sister group. Late Middle Cambrian and Upper Cambrian stages (at left) follow Ludvigsen and Westrop (1985). Base of the Marjumiid Biomere is placed at the base of the *Ehmaniella* Zone, following unpublished work of F. A. Sundberg.

stratigraphic context of these cladistic relationships alters the alleged turnover at biomere boundaries and the Cambrian-Ordovician boundary. The known Middle Cambrian occurrences of Anomocaridae and Ceratopygidae, for example, calibrate divergence of their sister group (first known from Late Cambrian asaphids). Taxa at more basal positions on the cladogram can be inferred as diverging from Anomocaridae and its sister group before the observed record of that clade in the medial Middle Cambrian of north China (Zhang and Jell 1987; Lower Marjuman Stage of Ludvigsen and Westrop

[1985]). Their stratigraphic first appearances in the Late Cambrian suggest ghost lineages of significant duration.

The lower boundary of the Ptychaspid Biomere is crossed by most asaphine families/superfamilies (fig. 5.5). Observed (T2) are Pterocephaliidae, Dikelokephalinidae, "saukiids" (Dikelocephalacea), Anomocaridae, Ceratoypgidae, and Asaphidae; cladistically inferred, based on stratigraphic first appearance of sister groups (T1), are Remopleuridacea and Cyclopygacea (i.e., Taihungshaniidae [Nileidae and Cyclopygidae] [Fortey 1981]). Monophyletic revision of the taxa reveals that the Cambrian-Ordovician boundary has less effect than tradition holds. Dikelocephalinids, kainellids (Remopleuridacea), "saukiids," ceratopygids, and asaphids are known from both systems. We can infer that the same applies to Cyclopygacea (based on the presence of their asaphid sister group in the early Late Cambrian, apparently as far back as the Australian Mindyallan Stage; Öpik 1967). That the cyclopygacean clade is first known from basal Ordovician Symphysurinae (following Fortey 1983) and Taihungshaniidae (Lu 1975) reveals a dramatic ghost lineage. Marjuman Dikelocephalacea ("saukiids") calibrate divergence of Remopleuridacea well beyond the observed first occurrence of the (ingroup) family Kainellidae in Sunwaptan strata (Shergold 1975). Only pterocephaliids and (with question) ptychaspidids appear to have last appearances in the boundary interval ("critical period" of Westrop [1989]). If poorly known *Curiaspis* Sdzuy 1955 is indeed a ptychaspidid, as suggested by Briggs et al. 1988, that family's range is also extended into the Tremadoc. Thus this large sample of Trilobita does not uphold extensive turnover at high taxonomic levels at biomere boundaries or the Cambrian-Ordovician boundary.

The latter is illustrated for an even larger number of families by Briggs et al. (1988:fig. 9.2). Based on a similarly styled compilation, Westrop (1989) states that 42% of shelf-occurring families in North America are of known last occurrence in the "critical period" at the top of the Ptychaspid Biomere. He has elegantly demonstrated that this minority of eliminated families is significantly more likely to be confined to shelf biofacies and endemic North American distribution.

Cladistic revision of Asaphina, along with the example to follow, suggests that conventional notions of trilobite turnover between Cambrian and Ordovician systems have been affected by taxonomic artifacts. This suggestion is hardly surprising, since traditional trilobite workers have upheld the reality of paraphyletic groups, have been influenced by belief in a Cambrian-Ordovician "event," or have generally worked exclusively on one or the other side of that systemic boundary. Furthermore, failure to account for ghost lineages (Norell, chapter 3 of this volume) in assessing taxonomic survivorship can significantly underestimate diversity and ages of origination of clades (see fig.

5.5). In total, ranges of ghost lineages are fully half those of known Cambrian taxon ranges for Asaphina. This disparity in stratigraphic first occurrence of sister groups indicates that preservational biases produce distorted diversification profiles if internal nodes of cladograms are not placed in a stratigraphic context (cf. Smith [1988] on preservational artifacts in the Cambrian echinoderm record).

Parsimony and "Cryptogenesis"

The notion that post-Cambrian trilobites are of undetermined (or undeterminable) relationship to Cambrian taxa has had an effect of exaggerating turnover at the boundary interval. A problem has been the autapomorphic nature of certain highly ranked post-Cambrian groups. The order Phacopida (fig. 5.6) is an example of an allegedly Ordovician-originating clade of uncertain sister group relations. A widely endorsed notion is that phacopides with a proparian facial suture are most closely related to another post-Cambrian group, Calymenina. However, character support for this grouping is vague (the sole positive character in the Treatise ordinal diagnosis [Henningsmoen 1959] is "known protaspides of similar type" [cf. Whittington 1954, 1957a,b]) and seems to have been influenced by stratigraphic lumping. Beecher (1897), to cite a classic example, regarded Calymenidae as Proparia, despite nonconformity with the group's diagnosis (presence of proparian facial sutures). Others have retained Calymenina in a paraphyletic order, Ptychopariida (e.g., as in the phenetic analysis of Lin [1988]), or allied it with other, nonphacopide taxa (Bergström 1973). Least informatively, it has been classified as a separate order (Hahn and Hahn 1975).

Characters from early ontogeny appear to be useful in filtering out some of these highly derived characters and in recognizing more general similarities shared by phacopides and other trilobite taxa. Notably, there is potential for homologizing particular larval exoskeletal spines. For example, Chatterton (1971) applied Whittington's (1956a) notation for odontopleuride spines to lichides and proetides. A separate nomenclature for glabellar and genal spines in Encrinuridae (see Edgecombe et al. 1988:fig. 2) masks homology with topological equivalents in these nonphacopide taxa (see also Ramsköld and Chatterton 1991).

This longstanding problem of high-level relationships has been approached using parsimony analysis. Consider the relationships of several taxa whose monophyly is well supported but whose sister-group relationships are ambiguous or unknown—in particular, Calymenina, proparian Phacopida (Cheirurina + Phacopina), and other "cryptogenetic" orders whose species diversity is predominantly post-Cambrian (Proetida, Lichida, and Odonto-

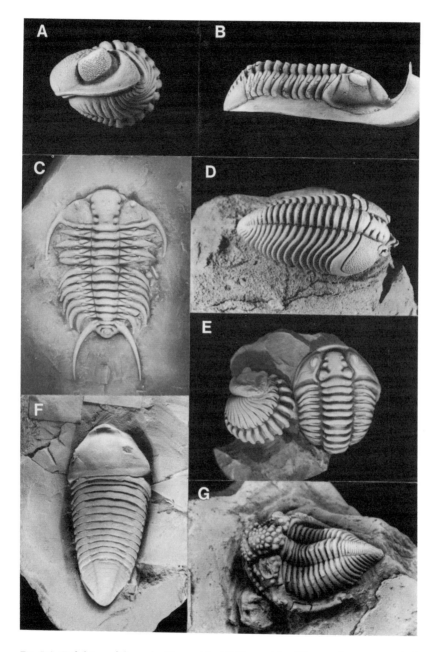

Fig. 5.6. Trilobites of the order Phacopida. (A) Phacopidae (*Phacops*; Devonian, ×1.4); (B) Acastacea, Acastidae (*Morocconites*; Devonian, ×0.9); (C) Cheiruridae (*Ceraurus*; Ordovician, ×1.0); (D) "Pliomeridae" (*Pseudocybele*; Ordovician, ×2.5); (E) Calymenidae (*Flexicalymene*; Ordovician, ×1.0); (F) Homalonotidae (*Trimerus*; Silurian, ×1.7); (G) Encrinuridae (*Erratencrinurus*; Ordovician, ×2.5).

pleurida). Twenty-seven characters of protaspid morphology were examined (tables 5.1 and 5.2) using the Early Cambrian "antagmine" *Crassifimbra* (Palmer 1958) as "ptychopariide" outgroup. This outgroup coding is consistent with the conventional hypothesis of generalized ptychopariide ancestry for cryptogenetic orders (e.g., Henningsmoen 1951:fig. 2; fig. 5.1). A recent phylogeny by Fortey (1990:fig. 19) would dispute monophyly of the analyti-

Table 5.1. Protaspid Exoskeletal Characters*

---

1. Cranidial anteromedian border: (0) absent; (1) present
2. Marginal spines on protocranidial anterior border: (0) absent; (1) sparse; (2) dense row
3. Preglabellar field: (0) absent; (1) present
4. Anterior axial furrow (fossular) pits: (0) large: (1) small or indistinct
5. Glabellar shape: (0) expanding forward; (1) parallel sided; (2) barrel shaped
6. Paired glabellar tubercles: (0) absent; (1) tubercles present; (2) coarse spines
7. Transglabellar furrows: (0) absent; (1) moderately to strongly incised
8. Frontal glabellar lobe with longitudinal furrows defining depressed abaxial region: (0) absent; (1) present
9. Isolated preoccipital lateral glabellar lobes: (0) absent; (1) present
10. Median occipital node: (0) absent; (1) present
11. Exoskeletal pitting: (0) absent; (1) present
12. Eye ridges: (0) present; (1) absent
13. Eye lobe swellings along anterior margin of protocranidium: (0) absent; (1) present.
14. Strongly proparian (librigena does not extend to posterior half of protocranidium): (0) present; (1) absent
15. Fixigenal spine (pf): (0) absent; (1) present; (2) with small subsidiary spine
16. Three large protocranidial marginal spines ("af," "mf," "pf" of Palmer [1962]): (0) absent; (1) present. Contra Fortey (1990: 565) the third protaspid spine pair in phacopide protaspides, "pf," is protocranidial and homologous with the genal spine of adults (cf. Chatterton et al. 1990: figs. 6.2, 6.3)
17. Arching of lateral margins: (0) gentle; (1) steep
18. Librigenal spine: (0) absent; (1) present
19. Dense row of librigenal-rostral border spines: (0) absent; (1) present
20. Number of marginal spines on hypostome (excluding posteromedian spine): (0) 0, 6, or 8
21. Elongate protopygidium (axis well in advance of posterior margin): (0) absent; (1) present
22. Paired axial spines on protopygidium: (0) absent; (1) present
23. Marginal spines on protopygidium: (0) absent; (1) present; (2) present, branching
24. Broad, flattened protopygidial pleural tips: (0) absent; (1) present
25. Protopygidial border furrow: (0) present; (1) absent
26. Protopygidial doublure (in section): (0) curved; (1) flat
27. Terrace lines on protopygidial doublure: (0) absent; (1) present

---

* See table 5.2 for distributions.

Table 5.2. Character Matrix for
Protaspid Exoskeletal Characters in Table 5.1

```
          11111111112222222
123456789012345678901234567
```
"Antagminae"
00000000000000000080000000
Lonchocephalidae
0000001000000100???00001??
Solenopleuropsinae
000000000000000010?10101??
Solenopleurinae
000000100100000?0???00001??
Proetida
10112100010100001001100010
Lichida
11011200110000000108111111
Odontopleurida
11011200110000200100111100
Calymenidae
12011110010010211018102010 0
Homalonotidae
12011110010010201018102010 0
Cheiruridae
01011110011011110006011010 0
Encrinuridae
11010110011011210016001010 1
Pterygometopidae
11010110011011211016001010 1
Phacopidae
1101011101?10120101?0010100
Dalmanitacea
01010111011001 1?101?00101??

cal ingroup used here. Fortey (1990) regards Phacopida as possibly nested within monophyletic Libristoma (a taxon nearly equivalent to "Ptychoparida," and including Proetida), but it depicts more distant common ancestry of Odontopleurida and Lichida. However, no explicit apomorphic features have been cited to support libristomate affinities for Phacopida. Phacopides lack the natent (i.e., not suturally attached) hypostome regarded as diagnostic of the group; reversal to a conterminant (attached) hypostome in Phacopida is forced by Fortey's hypothesis. Fortey's (1990) phylogeny would require the use of a redlichiid or olenellid outgroup if ingroup monophyly is to be maintained. It is noted that most of the larval apomorphies grouping odontopleurides and lichides with Phacopida (discussed later) are absent in these topologically primitive taxa, as well as Libristoma. Results retrieved with an early "ptychopariide" outgroup appear to be robust; plesiomorphic states are interpreted similarly with a nonlibristomate outgroup (e.g., the order Corynexochida).

Ambiguity (i.e., a poorly resolved consensus cladogram) remains when character data for these autapomorphic taxa are analyzed without additional "ptychopariides." This is analogous to considering relationships of solely Recent groups, to the exclusion of plesiomorphic fossil taxa. Particular character combinations of fossil taxa can affect topology and resolution (Donoghue et al. 1989; Novacek in chapter 2 of this volume). The traditional notion that generalized "ptychopariides" are the ancestral stock of most "cryptogenetic" orders (Henningsmoen 1951:fig. 2) warrants their inclusion in the data set. Considered are certain "ptychopariide" taxa suggested by previous investigators to be closest relatives of some part of the analytical ingroup. These include Late Cambrian Lonchocephalidae, suggested as grading into post-Cambrian Calymenidae (Briggs et al. 1988), and Solenopleuridae, regarded by some as near the ancestry of Calymenina and taxa here regarded as Proetida (Ahlberg and Bergström 1978:fig. 3). Protaspid morphology is afforded by the lonchocephalids *Welleraspis* and *Glaphyraspis* (see Chatterton et al. 1990 for species considered). Solenopleurids have been analyzed as two terminal taxa: Solenopleuropsinae (Middle Cambrian *Sao* [Whittington 1957a]) and Solenopleurinae (Middle Cambrian *Solenopleura* [Edgecombe in preparation]). This permits a test of the family's (sensu Poulsen 1955) monophyly; that Poulsen (p. 443) characterized this group as "numerous mediocre opisthoparian trilobites" suggests that the test is worthwhile.

Cladograms were constructed from the data in table 5.2 with the tree-building option ie★. With multistate characters nonadditive, 12 minimum-length cladograms (CI = 0.55) result in the strict consensus of figure 5.7. Larval characters provide support for some previously suggested groupings: proparian cheirurines and phacopines are a clade; monophyletic Phacopida includes Calymenidae and Homalonotidae (Calymenina); and Odontopleurida and Lichida are sister taxa (Chatterton 1980; Thomas and Holloway 1988). A novel hypothesis is the grouping of Proetida, Lichida + Odontopleurida, and Phacopida. Convention leads to the expectation that certain "ptychopariide" taxa would be found to be most closely related to ("ancestral to") each of these autapomorphic orders. Such is found to be an unparsimonious explanation of protaspid character distributions. The "ptychopariides" are excluded from the clade of "cryptogenetic" taxa. Apomorphic similarity between odontopleurides and phacopides finds support from detailed correspondences in distribution of genal border and epiborder spines (Ramsköld

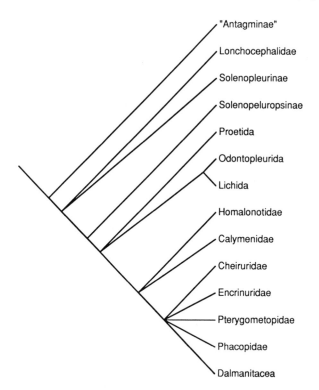

Fig. 5.7. Strict consensus of 12 minimum-length cladograms generated from protaspid character data in table 5.2. Consistency index (CI) is 0.55; retention index (RI), 0.67. Data analyzed with HENNIG 86, Version 1.5, using implicit enumeration (ie★).

and Chatterton 1991), for which homologs are not known in "ptychopariides"-libristomates.

None of the minimum-length cladograms unites "solenopleurids" as sister taxa; that taxon is evidently paraphyletic. Hypotheses such as lonchocephalid or solenopleurid ancestry for calymenids demand a decrease in cladogram consistency (see Ludvigsen et al. 1989 for a critique of supposed lonchocephalid-calymenid relations).

Successive approximations weighting (Carpenter 1988) with HENNIG 86 improves resolution over consensus methods, assigning higher weight to characters with greater cladistic reliability (sensu Farris 1969). Figure 5.8 depicts relationships in the minimum-length cladogram (CI = 0.79) generated with successive weighting; this affords more explicit predictions. Calymenina (Calymenidae + Homalonotidae) is supported, with a sister-group relation to

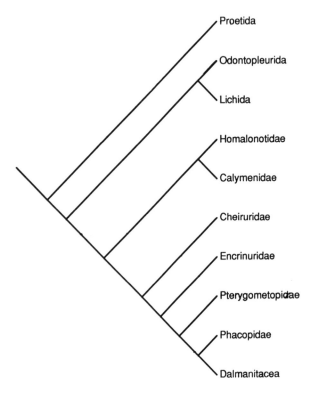

Fig. 5.8. Resolution of phacopide relationships (and two outgroup nodes) in minimum-length cladogram (CI = 0.79; RI = 0.87) generated from character data in table 5.2 with successive approximations weighting (xsteps w; ie* options of HENNIG 86). The suborders Calymenina and Phacopina are monophyletic; "Cheirurina" (Cheiruridae + Encrinuridae) is paraphyletic.

"Cheirurina" + Phacopina. The "cheirurine" grouping (Cheiruridae + Encrinuridae) is resolved as paraphyletic, encrinurids sharing more apomorphies with monophyletic Phacopina.

This conception of relationships of these orders permits inferences, of (minimum) age of origin (sensu Hennig), which alter the supposed context of "cryptogenesis" as a Cambrian-Ordovician boundary or Tremadocian "event." The case for alleged Late Cambrian Phacopida is addressed later (Calymenina and the proparian clade both have certain Tremadocian representatives). Late Middle Cambrian Lichida (Thomas and Holloway 1988) calibrate a minimum age for divergence from their odontopleuride sister group. This would imply a range extension from the observed first appearance of odontopleurides in the Late Cambrian (Bruton 1983); note, however, Ramsköld's (1991) suggestion that "Cambrian" odontopleurides may in fact be of Early Ordovician age. Shared ancestry of the odontopleuride-lichide clade with Phacopida is necessarily pre-late Middle Cambrian. Likewise this sets a minimum age of divergence (namely pre-late Middle Cambrian) for Proetida (Fortey and Owens 1975), despite the stratigraphic first appearance of this group near the Cambrian-Ordovician boundary ("Hystricurinae"). The implication is interesting that this clade of "cryptogenetic" orders has a separate history from supposed "ancestral ptychopariides" (e.g., lonchocephalids and "solenopleurids") back into Middle Cambrian. If divergence of these trilobite taxa from common ancestry is supposed to signal an Ordovician "event" (see fig. 5.1), that event now seems mythical. What remains warranted is the notion of marked species proliferation in these taxa during the Ordovician; the same cannot be said of their divergence.

Fortey's (1990:fig. 19) phylogeny includes even longer ghost lineages for certain clades. Lichida + Odontopleurida is removed from Libristoma-"Ptychopariida" and derived from a paraphyletic Lower Cambrian group, Redlichiida. This, of course, carries with it the inference of divergence from Libristoma before that group first occurs, in the Early Cambrian.

The "Ordovician Radiation": Taxonomic Congruence in Phacopida

Cladistic relationships of the order Phacopida afford a case study in the so-called Ordovician radiation, with correction for ghost lineages supplementing stratigraphic first appearance of terminal taxa. They also allow an empirical test of taxonomic congruence between larval and adult characters for trilobites. It has sometimes been argued that trilobite larval morphology is of dubious or limited value for phylogenetic inference. Bergström (1977:104), for example, cautioned that the "value [of trilobite larval characters] for phylogenetic discussion should not be exaggerated. After all, the larvae were not primarily phylogenetic recapitulatory stages but active and functioning animal stages with modifications of their own, no doubt partly different from those of adults. Similarly, Lane and Thomas (1983:152) claimed that a "particular morphology [in early ontogeny] presumably more immediately reflects a function for the individual and not any exact relationship between it and an individual belonging to another taxon at any particular stage in its development." Thomas and Holloway (1988:245) suggest that homoplasy should be rampant in trilobite larvae with similar life habits. Adaptationist suppositions imply that taxonomic incongruence is expected for phylogenetic hypotheses based on larval versus adult characters. This was opposed by

Hennig (1966), who clearly perceived that different stages of the life cycle (semaphoronts) will necessarily converge on a single pattern of genealogical relationships (cf. Brooks and Wiley 1985). Taxonomic congruence between large sets of larval and adult characters in Platyhelminthes has been demonstrated empirically by Brooks et al. (1985), without requiring "recapitulation" or recourse to adaptive scenarios.

The monophyly and ingroup relationships of Phacopida revealed by protaspides can be compared with those based on holaspid morphology. The latter have been inferred based on 40 exoskeletal characters (tables 5.3 and 5.4), coding for family-level taxa referred to the order. A full discussion of

Table 5.3. Holaspid Exoskeletal Characters *

1. Arching (tr.) of cranidial anterior margin: (0) flat or weakly arched medially; (1) strongly arched medially, with rolled doublure
2. Cranidial anterior border: (0) lengthening sagittally (anterior border furrow arched back); (1) equal length or shorter sagittally
3. Preglabellar field: (0) present; (1) absent
4. Glabellar expansion: (0) glabella parallel-sided or narrowing forward; (1) glabella broadening forward
5. Auxiliary impression system on frontal glabellar lobe: (0) indistinct; (1) ovate, circular; (2) triangular/rhomboid
6. Glabellar furrows nearly effaced: (0) absent (at least S1 well incised); (1) present
7. S4: (0) distinct, separate from S3; (1) indistinct; (2) merged with S3 exsagittally
8. Orientation of S3: (0) inclined forward sagittally; (1) inclined backward sagittally; (2) reduced to apodemal pits
9. Length (tr.) of S3: (0) short (<S2); (1) sinuous, equal to or longer than S2
10. Relative length (exsag.) of lateral glabellar lobes: (0) L1>L3; (1) L3>L1
11. Orientation of S2: (0) declined backward sagittally; (1) transverse; (2) inclined forward sagittally
12. S2 in contact with axial furrow: (0) present; (1) absent (L2–L3 coalesced abaxially)
13. Incision of S1 versus S2–S3: (0) subequal or grading in incision backward; (1) S1 much more strongly incised than S2–S3
14. Shape of S1: (0) evenly convex; (1) sigmoid
15. Course of S0: (0) transverse; (1) arched forward medially, occipital ring much shorter abaxially
16. Occipital spine: (0) node; (1) elongate spine with stout base; (2) absent
17. Eye type: (0) holochroal; (1) schizochroal; (2) blind
18. Eye ridges: (0) present; (1) absent
19. Circumocular sutures: (0) present; (1) absent
20. Genal pitting: (0) absent; (1) present
21. Granulate prosopon: (0) present; (1) absent
22. Muscle impression scar on fixigena near base of glabella: (0) absent; (1) present
23. Course of posterior branch of facial suture: (0) opisthoparian; (1) gonatoparian; (2) proparian; (3) fused

24. Rostral plate: (0) broadly trapezoidal; (1) narrow, tapering ventrally, with sinuous connective sutures; (2) absent
25. Doublure terrace lines: (0) present; (1) absent
26. Functional hypostomal suture: (0) absent (natent condition); (1) present
27. Hypostomal anterior wing process in association with anterior boss: (0) absent; (1) present
28. Maculae: (0) absent or indistinct; (1) distinct or prominent
29. Shouldering of hypostomal margin behind midlength (exsag.): (0) weak or absent; (1) prominent
30. Hypostomal marginal spines: (0) absent; (1) present
31. Length of hypostomal border (sag. vs. exsag.): (0) shorter or subequal sagittally; (1) elongated sagitally
32. Posteromedian embayment in hypostomal margin: (0) absent; (1) present
33. Swollen nodes on proximal part of thoracic pleurae: (0) absent; (1) present
34. Thoracic pleural articulating facet: (0) weak or absent (with free spine short or absent); (1) epifacetal; (2) absent (with free spine long)
35. Pygidial margin (0) entire; (1) with free pleural tips/spines
36. Number of pygidial pleurae: (0) five or fewer; (1) six or more
37. Number of pygidial axial rings-pleural ribs: (0) equal; (1) greater number of rings
38. Pseudoarticulating half-rings: (0) absent; (1) present
39. Pygidial pleural furrows: (0) distinct; (1) indistinct
40. Pygidial border: (0) present (flattened or rimlike); (1) absent

*See table 5.4 for distributions. 0 is plesiomorphic state; 1–3 are apomorphic states.

terminal taxa and codings will be presented elsewhere; a few explanatory remarks are necessary here. Families that appear to be paraphyletic are limited to the nominate subfamily (Pliomeridae; Pterygometopidae, see Ramsköld and Werdelin 1991). Encrinuridae includes Staurocephalidae and Dindymenidae to maintain its monophyly (Edgecombe et al. 1988); coding is based on primitive "Cybelinae." Pilekiidae Sdzuy 1955 is analytically treated as a separate family (cf. Whittington 1961; Jell 1985) to test cheirurid (Lane 1971) versus pliomerid (Harrington 1959) affinities. Experimental runs, in which a basal coding for the diverse and long-ranging family Cheiruridae was used (based on Cyrtometopinae-primitive Sphaerexochinae), grouped monotypic Hammatocnemidae (=*Ovalocephalus* Koroleva) with a pliomerine-encrinurid-Phacopina clade. However, when certain apomorphic ingroup cheirurids were also coded (e.g., the genus *Actinopeltis*), hammatocnemids nested within this cheirurid clade; exclusion of Hammatocnemidae as a separate family (Lu and Zhou 1979) renders Cheiruridae paraphyletic. Acastacea is sensu Eldredge (1979); coding is based on the Ordovician stem group "Kloucekiinae." Gyrometopinae (=*Gyrometopus* Jaanusson) and Diaphanometopinae (=*Diaphanometopus* Schmidt) were coded separately to test Jaanusson's (1975) hypothesis about their grouping. The family Echinophacopidae Zhou and Campbell 1990 is not used in this analysis; this monophyletic group is

Table 5.4. Holaspid Characters, with Alternative Character States as Described in Table 5.3

```
              1111111111222222222233333333334
    123456789012345678901234567890123456789 0
```

"Antagminae"
000000000000000000000000000000000000000000
Cliffiidae
0000001100000000?000000???????????010000
Olenidae
0100001100000100000010000000000000000000
Lonchocephalidae
0000001100000000100101000010000000000000
Calymenidae
1000001100000102000001101111100101000001
Homalonotidae
1000001100000102000001111110000101000001
Carmonidae
101000120??0000221?0000111?0100101000011
Eucalymenidae
1110001200000102?0?0000?0??????0?010001
Pilekiidae
0110002100000002001100201?10000002100101
Cheiruridae
0110001100000002001100201110000012100001
Hammatocnemidae
0111000001200012011100001?0000012100011
Pliomerinae
0111001101000012001100201110001001100011
Encrinuridae
0111002111100010001100201111011001101001
Diaphanometopinae
0110010000000020110?020????????01100001
Gyrometopinae
0111011111010012001100201???????01001101
Prosopiscidae
0110001111100002120?1003?????????01010100
Pterygometopidae
0111101111200012111100221111001001011001
Phacopidae
0111101111211012111000221110011001011101
Dalmanitacea
0111201111100101111002211010110010111 00
Acastacea
0111201111110121111002211010100101 11100
Actinopeltis
0111001001200002011100201 1?0000012100001

unambiguously nested within the cladistic structure of Phacopidae. *Bavarilla* Barrande is excluded from Homalonotidae (Fortey 1990).

"Ptychopariide" families suggested by previous investigators to be the sister group to some part of Phacopida were coded with the "antagmine" (Rasetti 1955) outgroup. These include Olenidae (allied with Cheirurina + Phacopina by Eldredge [1977:fig. 8]; coding is based on the primitive genus [Henningsmoen 1957] *Olenus*) and Lonchocephalidae (see earlier comments). Similarities between phacopides and Late Cambrian *Cliffia* Wilson 1951 were noted by Fortey (1990); relationship is assessed by coding the family Cliffiidae Hohensee (Hohensee and Stitt 1989) (cf. Phylacteridae Ludvigsen and Westrop in Ludvigsen et al. 1989). One might surmise that these "ptychopariides" provide a test for phacopide monophyly.

With multistate characters nonadditive, 248 minimum-length cladograms (CI = 0.49) are retrieved with the option 'mhennig★' and 'bb★' branch swapping. Figure 5.9 is the strict consensus of these topologies. Successive weighting applied to these cladograms selects eight of the minimum-length topologies (fig. 5.10). Phacopide monophyly is supported in all resolutions (cf. Fortey 1990), but unambiguous character support is rather scant. Ambiguity within the suborder Phacopina is especially reduced with rescaled character weights; multiple equally parsimonious resolutions result largely from autapomorphies in monotypic Prosopiscidae. Certain distinctive synapomorphies of pterygometopines, phacopids, and Dalmanitacea + Acastacea (schizochroal eyes, fused connective sutures) are represented by unique conditions in *Prosopiscus* (blindness, lack of dorsal sutures) such that the taxon's position must be evaluated on other characters. Cladograms with successive weighting (see fig. 5.10) corroborate the views that *Gyrometopus* is sister group to Phacopina (Jaanusson 1975), "pterygometopids" are most closely related to Phacopidae (Eldredge 1971), and Prosopiscidae is allied to Dalmanitacea + Acastacea (Fortey and Shergold 1984).

Taxonomic congruence between protaspid and holaspid data sets is high (fig. 5.11); both predict monophyly of Phacopida. Calymenine monophyly is retrieved from both analyses. The more taxonomically diverse adult data set indicates that the monotypic/low-diversity opisthoparian families Carmonidae (Kielan 1959) and Eucalymenidae (Lu 1975) are most closely related to traditional Calymenina (monophyletic Homalonotidae + Calymenidae). "Cheirurine" paraphyly is a prediction of both character sets. Holaspid morphology diagnoses Cheiruridae (including Hammatocnemidae) and groups Pilekiidae with it, based on structure of the thoracic pleurae. Phacopine monophyly is also indicated by both protaspid and holaspid characters. An incongruence in ingroup resolution (Pterygometopidae + Phacopidae versus Dalmanitacea + Phacopidae) might be an artifact of the dubious assignment

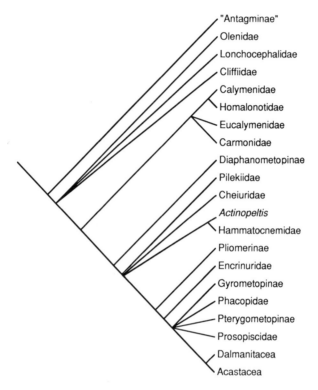

"Antagminae"
Olenidae
Lonchocephalidae
Cliffiidae
Calymenidae
Homalonotidae
Eucalymenidae
Carmonidae
Diaphanometopinae
Pilekiidae
Cheiuridae
Actinopeltis
Hammatocnemidae
Pliomerinae
Encrinuridae
Gyrometopinae
Phacopidae
Pterygometopinae
Prosopiscidae
Dalmanitacea
Acastacea

Fig. 5.9. Strict consensus of 248 minimium-length cladograms (CI = 0.49; RI = 0.69) based on holaspid character data from table 5.4. Cladograms were constructed using mhennig* bb* of HENNIG 86, with multistate characters nonadditive.

of protaspides to Phacopidae. Phacopid protaspid characters were coded based on Whittington's (1956b) Devonian larva (it is noted that protaspides assigned to *Phacops* by Chatterton [1971] are evidently those of a scutelluine; B. D. E. Chatterton personal communication). Association of Whittington's "phacopid" protaspis with mature growth stages is uncertain, and it could be that of a dalmanitid.

Figure 5.12 depicts the observed stratigraphic first appearance of phacopide terminal taxa from figure 5.10. Ghost lineages will be most significant if arguments for Late Cambrian "Cheirurina" (and their controversial dating) can be substantiated (Přibyl et al.'s (1985) classification of "Late Cambrian" *Emsurina* Sivov 1955 as a sphaerexochine cheirurid, *Emsurella* Rosova 1960 as a pliomerid, and *Eocheirurus* Rosova 1960 as a pilekiid would imply substantial differentiation of Phacopida already in the Cambrian (with range

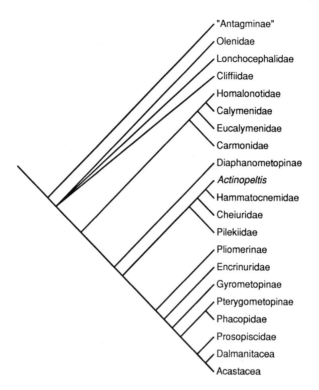

Fig. 5.10. Strict consensus of eight minimum-length cladograms (CI = 0.67; RI = 0.83) based on holaspid character data in table 5.4, generated with successive approximations weighting (xsteps w). Cladograms were constructed using mhennig* bb* of HENNIG 86, with multistate characters nonadditive.

extensions for Calymenina and Phacopina)). Until the Siberian taxa and their age relations are better known, this is suspect. Pilekiid affinities for *Eocheirurus* seem plausible (e.g., with its forked anterior glabellar furrow). *Emsurella*, however, is similar to the leiostegiacean family Cheilocephalidae (Westrop 1986:68).

Diversification histories with Cambrian pilekiids (see fig. 5.12A) contrast with a more cautious interpretation (see fig. 5.12B), which recognizes that divergence of a major phacopide clades had occurred within the Tremadoc. However, the only case of ghost lineages greater than a few instances of "Arenig" taxa with Tremadoc sister groups is within Calymenina. Carmonidae is known only from Late Ordovician strata; the hypothesis of relationships predicts its divergence from other calymenine clades (minimally) in the Tremadoc. In general, however, the interpretation of Phacopida as a post-

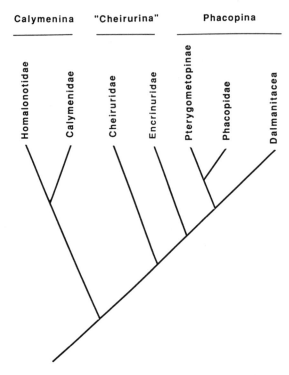

Fig. 5.11. Reduced cladogram derived from parsimony analysis of holaspid characters for Phacopida (see fig. 5.10). Relationships between phacopide families used in protaspid character analysis (see fig. 5.8) are depicted. Components are identical except within Phacopina (but see discussion in text).

Cambrian group (see fig. 5.12B) is consistent with traditional notions of "cryptogenesis" and Early Ordovician radiation. A sizable portion of the group (i.e., Encrinuridae + Phacopina) cannot be traced below the Arenig, and most families first appear within that series (exceptions are Phacopidae, first known from the Ashgill [*Sambremeusaspis* in Lespérance and Sheehan 1987], and Acastacea, of Llanvirn first appearance [*Kloucekia* in Hammann 1972]). Cladistic structure conforms to a Tremadoc–Arenig radiation.

A recurring theme of cladistic revisions (e.g., Fortey and Chatterton 1988) is that trilobites reveal less dramatic turnover between Cambrian and Ordovician than predicted by gradistic taxonomies. This finding is more pronounced in some taxa (e.g., Asaphina) than in others (e.g., Phacopida), where "cryptogenesis" persists. Evidence from larval characters suggests that diver-

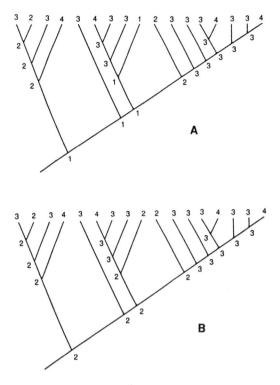

Fig. 5.12. Alternative stratigraphic interpretations of phacopide diversification. Cladogram topology is as shown in figure 5.10. Numerals at terminals refer to observed stratigraphic first appearance; states at internal nodes (minimum age of divergence) are inferred from the older of two sister taxa. 1, Latest Cambrian; 2, Tremadoc; 3, Arenig; 4, post-Arenig. (A) Late Cambrian taxa assigned to Pilekiinae (e.g., *Eocheirurus* Rosova); (B) Oldest ingroup taxa (Pilekiidae; Calymenidae s.l.; "Pliomeridae") of Tremadoc age.

gence of the "cryptogenetic" orders had occurred by the late Middle Cambrian, and separate "ptychopariide" ancestries are unparsimonious. Apparent high turnover of low-ranking taxa at certain intervals is especially pronounced in platform settings (these are the biomere boundaries). However, correcting for paraphyly and ghost lineages, the notion of familial and ordinal groups revealing distinct Cambrian and Ordovician "phases of diversification" is diminished.

The same applies to models of biomeres as monophyletic radiations following mass extinctions. Durations of Cambrian ghost lineages, a measure of preservational incompleteness, are significant in accounting for taxonomic range. Perhaps, as with echinoderms (Smith 1988), a single "phase" of Lower Paleozoic trilobite diversification may be most compatible with cladistic patterns.

ACKNOWLEDGMENTS

This chapter has drawn from progress in trilobite cladistics and the Cambrian-Ordovician boundary made by Richard Fortey and Steve Westrop. Brian Chatterton compiled the first version of our ordinal protaspid data set. Mark Norell is to blame for piquing my interest in ghost lineages and the like. The manuscript has been improved by comments from Gareth Nelson, as well as those of my Cambrian colleagues Fred Sundberg and Steve Westrop. I thank Niles Eldredge for getting me involved in the Cornell symposium and for discussion on things trilobed.

REFERENCES

Ahlberg, P. and J. Bergström. 1978. Lower Cambrian ptychopariid trilobites from Scandinavia. *Sveriges Geol. Undersök.* 49:1–41.

Ax, P. 1987. *The Phylogenetic System: The Systematization of Organisms on the Basis of Their Phylogenesis.* Chichester: John Wiley.

Barrande, J. 1846. Notice préliminaire sur le Silurien et les Trilobites de Bohême. Leipzig.

Beecher, C. E. 1897. Outline of a natural classification of the trilobites. *Amer. J. Sci.* 3:89–106, 181–207.

Bergström, J. 1973. Organization, life, and systematics of trilobites. *Fossils and Strata* 2:1–69.

Bergström, J. 1977. Proetida—a disorderly order of trilobites. *Lethaia* 10:95–105.

Briggs, D. E. G., R. A. Fortey, and E. N. K. Clarkson. 1988. Extinction and the fossil record of the arthropods. In G. P. Larwood, ed., *Extinction and Survival in the Fossil Record,* pp. 171–209. Oxford: Clarendon Press.

Brooks, D. R., R. T. O'Grady, and D. R. Glen. 1985. Phylogenetic analysis of the Digenea (Platyhelminthes: Cercomeria) with comments on their adaptive radiation. *Can. J. Zool.* 63:411–443.

Brooks, D. R. and E. O. Wiley. 1985. Theories and methods in different approaches to phylogenetic systematics. *Cladistics* 1:1–11.

Bruton, D. 1983. Cambrian origins of the odontopleurid trilobites. *Palaeontology* 26:875–885.

Carpenter, J. M. 1988. Choosing among multiple equally parsimonious cladograms. *Cladistics* 4:291–296.

Chatterton, B. D. E. 1971. Taxonomy and ontogeny of Siluro-Devonian trilobites from near Yass, New South Wales. *Palaeontographica A* 137:1–108.

Chatterton, B. D. E. 1980. Ontogenetic studies of Middle Ordovician trilobites from the Esbataottine Formation, Mackenzie Mountains, Canada. *Palaeontographica A* 171:1–74.

Chatterton, B. D. E., D. J. Siveter, G. D. Edgecombe, and A. S. Hunt. 1990. Larvae and relationships of the Calymenina. *J. Paleontol.* 64:255–277.

Donoghue, M. J., J. A. Doyle, J. Gauthier, A. G. Kluge, and T. Rowe. 1989. The importance of fossils in phylogeny reconstruction. *Ann. Rev. Ecol. Syst.* 20:431–460.

Edgecombe, G. D. 1991. Phyla, phylogeny, and fantasy. *Cladistics* 7:101–103.

Edgecombe, G. D., S. E. Speyer, and B. D. E. Chatterton. 1988. Protaspid larvae and phylogenetics of encrinurid trilobites. *J. Paleontol.* 62:779–799.

Eldredge, N. 1971. Patterns of cephalic musculature in the Phacopina (Trilobita) and their phylogenetic significance. *J. Paleontol.* 45:52–67.

Eldredge, N. 1977. Trilobites and evolutionary patterns. In A. Hallam, ed., *Patterns of Evolution as Illustrated by the Fossil Record*, pp. 205–332. Amsterdam: Elsevier.

Eldredge, N. 1979. Cladism and common sense. In J. Cracraft and N. Eldredge, eds., *Phylogenetic Analysis and Paleontology*, pp. 165–198. New York: Columbia University Press.

Farris, J. S. 1969. A successive approximations approach to character weighting. *Syst. Zool.* 18:374–385.

Farris, J. S. 1988. HENNIG 86. Version 1.5. Distributed by the author, 41 Admiral Street, Port Jefferson Station, N.Y.

Flessa, K. W. and D. Jablonski. 1983. Extinction is here to stay. *Paleobiology* 9:315–321.

Foote, M. 1989. Taxon-free analysis of trilobite morphospace. *Geol. Soc. Amer., Ann. Meeting, Abstr. Prog.* 21:288.

Fortey, R. A. 1981. *Prospectatrix genatenta* (Stubblefield) and the trilobite superfamily Cyclopygacea. *Geol. Mag.* 118:603–614.

Fortey, R. A. 1983. Cambrian-Ordovician trilobites from the boundary beds in western Newfoundland and their phylogenetic significance. In D. E. G. Briggs and P. D. Lane, eds., *Trilobites and Other Early Arthropods*, pp. 179–211. Special Papers in Palaeontology, 30. London: The Paleontological Association.

Fortey, R. A. 1989. There are extinctions and extinctions: examples from the Lower Paleozoic. *Phil. Trans. R. Soc. Lond. B* 325:327–355.

Fortey, R. A. 1990. Ontogeny, hypostome attachment, and the classification of trilobites. *Palaeontology* 33:529–576.

Fortey, R. A. and B. D. E. Chatterton. 1988. Classification of the trilobite suborder Asaphina. *Palaeontology* 31:165–222.

Fortey, R. A. and R. M. Owens. 1975. Proetida—a new order of trilobites. *Fossils and Strata* 4:227–239.

Fortey, R. A. and J. H. Shergold. 1984. Early Ordovician trilobites, Nora Formation, Central Australia. *Paleontology* 27:315–366.

Hahn, G. and R. Hahn. 1975. Forschungbericht über Trilobitomorpha. *Paläont. Z.* 49:432–460.

Hammann, W. 1972. Neue propare Trilobiten aus dem Ordovizium Spaniens. *Senckenbergiana lethaea* 53:371–381.

Hardy, M. 1985. Testing for adaptive radiation: the Ptychaspid (Trilobita) Biomere of the Cambrian. In J. W. Valentine, ed., *Phanerozoic Diversity Patterns: Profiles in Macroevolution*, pp. 379–397. Princeton: Princeton University Press.

Harrington, H. J. 1959. Pliomeridae. In R. C. Moore, ed., *Treatise on Invertebrate Paleontology. Part O. Arthropoda 1*, pp. 439–445. Lawrence: Geological Society of America, University of Kansas Press.

Henderson, R. A. 1976. Upper Cambrian (Idamean) trilobites from western Queensland, Australia. *Palaeontology* 19:325–364.

Hennig, W. 1965. Phylogenetic Systematics. *Ann. Rev. Entomol.* 10:97–116.

Hennig, W. 1966. *Phylogenetic Systematics*. Urbana: University of Illinois Press.

Hennig, W. 1969. *Die Stammesgeschichte der Insekten*. Frankfurt am Main: E. Kramer.

Henningsmoen, G. 1951. Remarks on the classification of trilobites. *Norsk Geol. Tidsskr.* 29:174–217.

Henningsmoen, G. 1957. The trilobite family Olenidae with description of Norwegian material and remarks on the Olenid and Tremadocian Series. *Skrift. Norske Vidensk. Akad. Oslo I. Matemat. Naturvidensk. Klasse* 1957 1:1–303.

Henningsmoen, G. 1959. Phacopida. In R. C. Moore, ed., *Treatise on Invertebrate Paleontology. Part O. Arthropoda 1*, p. 430. Lawrence: Geological Society of America, University of Kansas Press.

Henningsmoen, G. 1973. The Cambro-Ordovician boundary. *Lethaia* 6:423–439.

Hohensee, S. R. and J. H. Stitt. 1989. Redeposited *Elvinia* Zone (Upper Cambrian) trilobites from the Collier Shale, Ouachita Mountains, west-central Arkansas. *J. Paleontol.* 63:857–879.

Jaanusson, V. 1975. Evolutionary processes leading to the trilobite suborder Phacopina. *Fossils and Strata* 4:209–218.

Jefferies, R. P. S. 1979. The origin of chordates—a methodological essay. In M. R. House, ed., *The Origin of Major Invertebrate Groups*, pp. 443–477. London: Academic Press.

Jell, P. A. 1985. Tremadoc trilobites of the Digger Island Formation, Waratah Bay, Victoria. *Mem. Victoria Mus.* 46:53–88.

Kielan, Z. 1959. Upper Ordovician trilobites from Poland and some related forms from Bohemia and Scandinavia. *Palaeontol. Polon.* 11:1–198.

Lane, P. D. 1971. British Cheiruridae (Trilobita). *Palaeontogr. Soc. Monogr.* 125:1–95.

Lane, P. D. and A. T. Thomas. 1983. A review of the trilobite suborder Scutelluina. In D. E. G. Briggs and P. D. Lane, eds., *Trilobites and Other Early Arthropods*, pp. 141–160. Special Papers in Palaeontology, 30. London: The Palaeontological Association.

Lespérance, P. J. and P. M. Sheehan. 1987. Trilobites et Brachiopodes ashgilliens (Ordovicien supérieur) de l' "Assie" de Fosse, Bande de Sambre-Meuse (Belgique). *Bull. Inst. R. Sci. Nat. Belg., Sci. Terre* 57:91–123.

Lin, T. 1988. Application of cluster analysis to the taxonomy of order and suborder of the Trilobita. *Sci. Sinica. Ser. B* 31:1274–1280.

Lochman, C. 1956. The evolution of some Upper Cambrian and Lower Ordovician trilobite families. *J. Paleontol.* 30:445–463.

Longacre, S. A. 1970. Trilobites of the Upper Cambrian Ptychaspid Biomere, Wilberns Formation, Central Texas. *J. Paleontol. Mem.* 4:1–70.

Lu, Y.-H. 1975. Ordovician trilobite faunas of central and southwestern China. *Palaeontol. Sinica 152, Ser. B* 11:1–463.

Lu, Y.-H. and Z.-Y. Zhou. 1979. Systematic position and phylogeny of *Hammatocnemis* (Trilobita). *Acta Palaeontol. Sinica* 18:415–433.

Ludvigsen, R. 1982. Upper Cambrian and Lower Ordovician trilobite biostratigraphy of the Rabbitkettle Formation, western District of Mackenzie. *R. Ontario Mus., Life Sci. Contrib.* 134:1–188.

Ludvigsen, R. and S. R. Westrop. 1983. Franconian trilobites of New York State. *N.Y. State Mus. Mem.* 23:1–83.

Ludvigsen, R. and S. R. Westrop. 1985. Three new Upper Cambrian stages for North America. *Geology* 13:139–143.

Ludvigsen, R., S. R. Westrop, and C. H. Kindle. 1989. Sunwaptan (Upper Cambrian) trilobites of the Cow Head Group, western Newfoundland, Canada. *Palaeontogr. Can.* 6:1–175.

Nelson, G. and N. Platnick. 1981. *Systematics and Biogeography. Cladistics and Vicariance.* New York: Columbia University Press.

Norford, B. S. 1988. Introduction to papers on the Cambrian-Ordovician boundary. *Geol. Mag.* 125:323–326.

Öpik, A. A. 1967. The Mindyallan fauna of North-western Queensland. *Bur. Miner. Res. Geol. Geophys. Bull.* 74:1–404.

Palmer, A. R. 1958. Morphology and ontogeny of a Lower Cambrian ptychoparioid trilobite from Nevada. *J. Paleontol.* 32:154–170.

Palmer, A. R. 1962. Comparative ontogeny of some opisthoparian, gonatoparian and proparian Upper Cambrian trilobites. *J. Paleontol.* 36:87–96.

Palmer, A. R. 1965a. Biomere—a new kind of biostratigraphic unit. *J. Paleontol.* 39:149–153.

Palmer, A. R. 1965b. Trilobites of the Late Cambrian Pterocephaliid Biomere in the Great Basin, United States. U.S. Geological Survey, Professional Paper 493:1–105.

Palmer, A. R. 1981. Subdivision of the Sauk Sequence. In M. E. Taylor, ed., *Short Papers for the Second International Symposium on the Cambrian System. 1981,* pp. 160–162. U.S. Geological Survey, Open-File Report 81–743.

Palmer, A. R. 1984. The biomere problem: evolution of an idea. *J. Paleontol.* 58:599–611.

Platnick, N. I. 1985. Philosophy and the transformation of cladistics revisited. *Cladistics* 1:87–94.

Poulsen, C. 1955. Attempt at a classification of the trilobite family Solenopleuridae. *Medel. Dansk Geol. Foren.* 12:443–447.

Přibyl, A., J. Vaněk, and I. Pek. 1985. Phylogeny and taxonomy of family Cheiruridae (Trilobita). *Acta Univ. Palack. Olomuc. Facul. Rer. Natur. Geogr. Geol.* 83:107–193.

Ramsköld, L. 1991. Pattern and process in the evolution of the Odontopleuridae (Trilobita). The Solenopeltinae and Ceratocephalinae. *Trans. R. Soc. Edinburgh. Earth Sci.* 82:143–181.

Ramsköld, L., and B. D. E. Chatterton. 1991. Revision of the polyphyletic *"Leonaspis"* (Trilobita). *Trans. R. Soc. Edinb. Earth Sci.* 82.

Ramsköld, L., and L. Werdelin. 1991. Phylogeny and evolution of some phacopid trilobites. *Cladistics* 7:29–74.

Rasetti, F. 1955. Lower Cambrian ptychopariid trilobites from the conglomerates of Quebec. *Smithsonian Misc. Collect.* 128:1–35.

Rieppel, O. 1988. *Fundamentals of Comparative Biology.* Basel: Birkhäuser Verlag.

Rosova, A. V. 1960. Trilobity iz otlozenij tolstocichinskoj svity Salaira. *Tr. Sibirsk Nauch. Issled. Inst. Geol. Geofys. Miner.* 5:1–116.

Sdzuy, K. 1955. Die fauna der Leimitz-Schiefer (Tremadoc). *Abh. Senckenberg. Naturforsch. Gesell.* 492:1–74.

Sepkoski, J. J., Jr. 1979. A kinetic model of Phanerozoic taxonomic diversity II. Early Phanerozoic families and multiple equilibria. *Paleobiology* 5:222–251.

Sepkoski, J. J., Jr. 1981. The uniqueness of the Cambrian fauna. In M. E. Taylor, ed., *Short Papers for the Second International Symposium on the Cambrian System. 1981,* pp. 203–207. U.S. Geological Survey, Open File Report 81–743.

Sepkoski, J. J., Jr. 1986. Phanerozoic overview of mass extinction. In D. M. Raup and D. Jablonski, eds., *Patterns and Processes in the History of Life,* pp. 277–295. Berlin: Springer-Verlag.

Shergold, J. H. 1975. Late Cambrian and Early Ordovician trilobites from the Burke River structural belt, Western Queensland, Australia. *Bur. Miner. Res. Geol. Geophys. Bull.* 153:1–251.

Sivov, A. G. 1955. New genera of trilobites. In L. L. Khalfin, ed., *Atlas of Guide Forms of Fossil Faunas and Floras of Western Siberia,* vol. 1. Tomsk: Western Siberian Geological Institute.

Smith, A. B. 1988. Patterns of diversification and extinction in Early Paleozoic echinoderms. *Palaeontology* 31:799–828.

Smith, A. B. and C. Patterson. 1988. The influence of taxonomic method on the perception of patterns of evolution. *Evol. Biol.* 23:127–216.

Speyer, S. E. and B. D. E. Chatterton. 1989. Trilobite larvae and larval ecology. *Hist. Biol.* 3:27–60.

Stanley, S. M. 1979. *Macroevolution: Pattern and Process.* San Francisco: W. H. Freeman.

Stitt, J. H. 1971. Repeating evolutionary pattern in Late Cambrian trilobite biomeres. *J. Paleontol.* 45:178–181.

Stitt, J. H. 1975. Adaptive radiation, trilobite paleoecology, and extinction, Ptychaspidid Biomere, Late Cambrian of Oklahoma. *Fossils and Strata* 4:381–390.

Stitt, J. H. 1977. Late Cambrian and earliest Ordovician trilobites, Wichita Mountains area, Oklahoma. *Okla. Geol. Surv. Bull.* 124:1–79.

Stitt, J. H. 1983. Trilobites, biostratigraphy, and lithostratigraphy of the Mackenzie Hill Limestone (Lower Ordovician), Wichita and Arbuckle Mountains, Oklahoma. *Okla. Geol. Surv., Bull.* 134:1–54.

Stubblefield, C. J. 1959. Evolution in trilobites. *Q. J. Geol. Soc. Lond.* 115:145–162.

Sundberg, F. A. 1989. Morphological diversification of trilobites within Cambrian biomeres. *Geol. Soc. Amer., Ann. Meeting, Abstr. Prog.* 21:288.

Thomas, A. T. and D. J. Holloway. 1988. Classification and phylogeny of the trilobite order Lichida. *Phil. Trans. R. Soc. Lond. B. Biol. Sci.* 321:179–262.

Westrop, S. R. 1986. Trilobites of the Upper Cambrian Sunwaptan Stage, southern Canadian Rocky Mountains. *Palaeontogr. Can.* 3:1–179.

Westrop, S. R. 1989. Trilobite mass extinction near the Cambrian-Ordovician boundary in North America. In S. K. Donovan, ed., *Mass Extinction: Processes and Evidence*, pp. 89–103. London: Belhaven Press.

Westrop, S. R. 1990. Mass extinction in the Cambrian trilobite faunas of North America. In S. J. Culver, ed., *Arthropod Paleobiology*, pp. 99–115. Knoxville: University of Tennessee Studies in Geology.

Westrop, S. R. and R. Ludvigsen. 1987. Biogeographic control of trilobite mass extinction at an Upper Cambrian "biomere" boundary. *Paleobiology* 13:84–99.

Whittington, H. B. 1954. Status of Invertebrate Paleontology, 1953. VI. Arthropoda: Trilobita. *Bull. Mus. Comp. Zool.* 112:193–200.

Whittington, H. B. 1956a. Silicified Middle Ordovician trilobites: the Odontopleuridae. *Bull. Mus. Comp. Zool.* 114:155–288.

Whittington, H. B. 1956b. Beecher's supposed odontopleurid protaspis is a phacopid. *J. Paleontol.* 30:104–109.

Whittington, H. B. 1957a. Ontogeny of *Elliptocephala, Paradoxides, Sao, Blainia,* and *Triarthrus* (Trilobita). *J. Paleontol.* 31:934–946.

Whittington, H. B. 1957b. The ontogeny of trilobites. *Biol. Rev.* 32:421–469.

Whittington, H. B. 1961. Middle Ordovician Pliomeridae (Trilobita) from Nevada, New York, Quebec, Newfoundland. *J. Paleontol.* 35:911–922.

Wilson, J. L. 1951. Franconian trilobites of the central Appalachians. *J. Paleontol.* 25:617–654.

Zhang, W. and P. A. Jell. 1987. *Cambrian Trilobites of North China. Chinese Cambrian Trilobites Housed in the Smithsonian Institution.* Beijing: Science Press.

Zhou, Z.-Q. and K. S. W. Campbell. 1990. Devonian phacopacean trilobites from the Zhusilenghaierhan region, Ejin Qi, western Inner Mongolia, China. *Palaeontographica A* 214:57–77.

6 : Vicariance Biogeography, Geographic Extinctions, and the North American Oligocene Tsetse Flies

David A. Grimaldi

Abstract. Fossil taxa occurring in areas remote from the closest living relatives can be the most robust test of vicariant biogeographic hypotheses, but only if the fossil taxon is not plesiomorphic to or the sister group of the living relatives. Cladistic criteria and conditions are proposed for analyzing the historical geographical significance of a fossil where its relationships are known. Several fossil arthropod examples are discussed where geographic extinctions have apparently been of major influence in biogeography, including the famous case of the tsetse flies from the Oligocene shales of Florissant, Colorado.

Eight of the nine known specimens of fossil tsetse flies (Diptera: Glossinidae) have been compared in detail for the first time. Two species, not four, exist, with the following synonyms proposed: *Glossina oligocenus* (Scudder) (=*G. veterna* Cockerell), *G. osborni* Cockerell (=*G. armatipes* Cockerell). Species characters are based on size and proportions in wing venation. The two Oligocene *Glossina* species are closely related, but plesiomorphic to the monophyletic group of living, Afrotropical *Glossina*, as based on two newly described characters of the legs, and possibly the absence of the characteristic tertiary-branched arista found in living *Glossina*. Thus the primitive North American fossil tsetses have little relevance to the biogeography of living Afrotropical species, other than establishing their minimum age as Upper Oligocene.

Vicariance Biogeography and Geographical Extinctions

Vicariance biogeography is an analytic technique used to discern similar patterns of branching among organisms and the land areas that they inhabit (for review, see Brundin 1981; Platnick and Nelson 1978; Humphries and Parenti 1986). If there is good concordance between the phylogeny of the taxa and the areas they inhabit, then presumably present-day distributions are due to geologic events that vicariated the organisms and their areas (if not, dispersal has been important).

Humphries and Parenti (1986) stated that there is "no real role for fossils in cladistic biogeography except to help in rejecting a geological explanation for a particular pattern." Brundin was of a similar philosophy: "Fossils will on the whole either confirm or complete the reconstructed patterns [based on living taxa], which has been the case so far with the chironomids. But the history of the extinct stem groups of Chironomidae and other groups of the same or greater age will largely remain lost because of insufficient documentation. The history of many typical relict groups, is also out of reach" (1990:368). It is my assertion, argued later, that a fossil can be used not only to reject a hypothesis of geological event(s) responsible for present distributions, but to modify a vicariance hypothesis. The criteria required for such modification are stringent, and necessarily limit the utility of fossils for biogeography, but if the phylogenetic and biogeographic position of a fossil indicates that the group in question at one time had a much wider distribution, then drawing vicariance hypotheses within the limits of a contracted, present distribution is obviously of limited value. Patterson (1981) was more generous in his recognition of the role of fossils in paleobiogeography, in which he stated, there are two roles: (1) a fossil documents phenotypic and/or geographic extinctions, and thus allows new test groups and areas to be considered in an analysis; (2) the minimum-age data of a fossil permit one to narrow the range of possible influential geologic events (e.g., if an Eocene fossil is found for a group hypothesized to have been affected only by Pleistocene events, that hypothesis can be rejected). The great amount of discussion on extinction has almost exclusively centered on taxon extinctions (e.g., Marshall 1990). The implications for extinctions on a geographic scale have rarely been explored (but see Grande 1985, discussed later), or the concern has been with ecological factors of range contraction or expansion (Vermeij 1989).

Patterson (1981) and in particular Grande (1985) have best explored geographic extinctions. Grande summarized the utility of fossil data for biogeography in four points, which are similar to those discussed by Patterson. His point 2 agrees most with geographic extinctions, that "fossils provide additional taxa which can increase the known biogeographic range of a taxon" (1985:236). Grande also showed how a good fossil record can help our understanding of a distribution pattern complicated by recent events, as based on fish faunas presently and formerly in North America. The crucial importance of phylogeny in placing fossil taxa was discussed, but not with reference to geographical extinctions.

Figure 6.1A–H shows eight combinations of phylogenetic position and geographic location of a fossil with respect to three living relatives. Concerning taxon A + B + C, examples A–C and E in figure 6.1 are less interesting than the others because the fossil locality lies in the midst of the present-day distribution. However, examples B, C, and E are still informative, since they at least indicate a minimum age for the area of endemism of taxon A + B + C. Situation A is less informative in this regard, since it is the sister group to A + B + C; but since sister groups, by cladistic definition, are of equivalent ages, one could reasonably suppose that the age of area of endemism of A + B + C is the same (minimum) age of the fossil. The major drawback of this supposition is the scarce fossil record of many groups: in situation A, any number of younger fossils might be discovered that are more closely related to living members than is the present fossil. In situations B, C, and E, younger fossils within the area of endemism would have little effect on dating the age of the area.

Examples D and G–H are instances where the fossil lies outside the

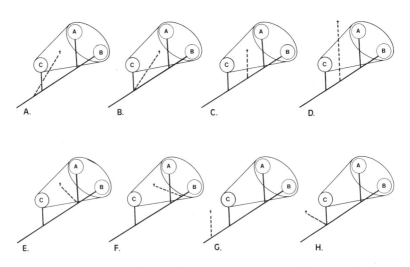

A.   B.   C.   D.

E.   F.   G.   H.

Fig. 6.1. Eight possibilities for the phylogenetic position and biogeographic location of a fossil (cross) with respect to its three closest living relatives (A, B, and C).

distribution of living relatives, and thus would seem important in being able to modify or refute biogeographic hypotheses based solely on taxa A, B, and C and the area they inhabit. Examples G and H are actually uninformative in this regard since the fossil is the sister group to A + B + C (in G) or one of three sister groups (in H). To conclude on the basis of examples G or H that group A + B + C had a more widespread distribution is simply redefining the clade to include another, plesiomorphic taxon (the fossil), and this approach can be used to encompass any number and combination of outgroups, living and extinct. Example G is the situation seen in the fossil tsetse flies, described later. In example D the fossil is the sister group to A + B, and it lies outside the distribution of A + B and C. Since the plesiomorphic state here for the distribution of taxon (A + B + fossil) lies somewhere *between* C and A + B, one can only conclude the position of the fossil is derived. Thus, in this example, a hypothesis of vicariance between areas A + B and C must be modified by the fossil evidence: taxon A + B + C formerly had a more widespread distribution. To exclude the fossil taxon would be defining the vicariant pattern of a paraphyletic group. Example F is similar to example D, except that it applies only with respect to vicariance hypotheses of taxa A and B. Among the eight examples, the one where fossil data can revise hypotheses of vicariance based on the living taxa, A–C, are examples D and F. How these hypothetical examples apply to actual organisms and data is explained later.

The oldest fossil bee, *Trigona prisca,* has recently been discovered in late Cretaceous amber from New Jersey (Michener and Grimaldi 1988). The find not only extends the geological record of the bees, but also adds another record, which would indicate that the stingless bees (Meliponinae: Apidae), to which *Trigona* belongs, formerly had a more widespread distribution (fig. 6.2). The group presently is circumtropical, but amber fossils are known as well from the Baltic, Sicily, and the Dominican Republic (no native stingless bees exist on the Antilles), well outside the distribution. Unfortunately, we have no cladistic analysis of the meliponines, living or extinct. If the amber fossils are all plesiomorphic with respect to the living meliponines, this situation is just as in example G, described earlier, and as in the case of the tsetse flies. Wille (1977) indicated that the Old World fossils are indeed plesiomorphic to the living meliponines, but others (e.g., *Trigona dominicana* [Michener, 1982] and *T. prisca*) are not plesiomorphic meliponines. Hypothesizing vicariant processes without the knowledge of *T. dominicana* and *T. prisca* would be misleading, for the group, at least in the New World, formerly had a more widespread distribution.

In the tropics of the Old World is a group of very bizarre diopsids with eyes that are barely stalked, genus *Sphyracephala*. This genus is more widespread in the Old World, but it also occurs in eastern North America (fig.

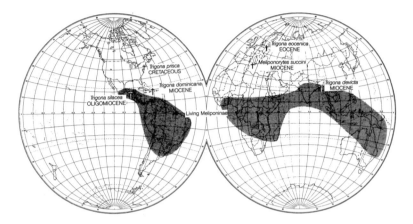

Fig. 6.2. Distribution of living and fossil stingless bees (Meliponinae). (From Wille 1977; Michener and Grimaldi 1988.)

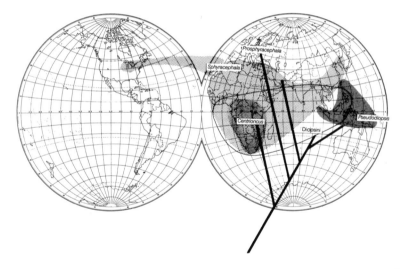

Fig. 6.3. Distribution of living and fossil Diopsidae (Diptera). [From Hennig (1965).]

6.3). The sister group to all the diopsids (stalk-eyed and barely so) is *Centrioncus* (not stalk-eyed at all), throughout Africa south of about 10° N latitude. Two fossils lie outside the distribution of living Diopsidae: one of them, *Prosphyracephala,* occurs in Baltic (Eocene to Miocene) amber. (The other diopsid fossil is from Oligocene rocks in Montana; it is too poorly preserved to know anything of its relationships within the diopsids.) Hennig (1965) placed *Prosphyracephala* as the sister group to the rest of the diopsids, sans *Centrioncus*. The primitive state of diopsid distribution thus being south African, this must mean that the entire family Diopsidae had a distribution that encompassed the Baltic amber at least during Baltic amber times. Excluding *Prosphyracephala* from cladistic analysis would be basing hypotheses of diopsid vicariance upon a paraphyletic group.

An example of a prototype austral distribution is shown by the living spider genera in the family Archaeidae (Forster and Platnick, 1984) (fig. 6.4). The Australian genus *Austrarchaea* is apparently the sister group to the

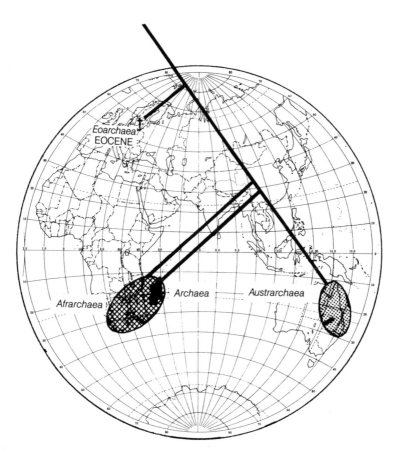

Fig. 6.4. Distribution of living and fossil Archaeidae (Araneae). (From Forster and Platnick 1984).

Malagasy genus *Archaea*. *Afrarchaea*, from southern Africa and Madagascar, is the sister group to these two genera. Based on just these three genera, one might hypothesize a vicariance event between Madagascar and southern Africa, and this joint area with Australia. These land masses, with Patagonian South America, have long been known to harbor organisms more closely related to each other than to organisms in northerly areas of their respective continents, as elegantly shown by Brundin (1966) with chironomid midges. *Eoarchaea*, in the Baltic (presumably Eocene) amber, is the sister group to the living archaeid spiders. As such, *Eoarchaea* has no effect on vicariance hypotheses of living archaeids, other than contributing a minimum age to this group by sister-group dating. This situation is very similar to that of the tsetse flies, discussed later. Interestingly, the chironomid midges are extremely common and diverse in the Baltic amber, yet, to my knowledge, Brundin and others have not restudied this wealth of material described by Meunier at the turn of the century. The Baltic amber chironomids could contribute some very valuable data on Brundin's hypotheses of austral vicariance in this group.

The Sciadoceridae are a group of small, obscure, yet distinctive Diptera, also with a present-day austral distribution. They are the sister group to the large extant family of scuttle flies, the Phoridae. The two living species are *Archiphora patagonica*, in southern Chile, and *Sciadocera rufomaculata*, in southeastern Australia and New Zealand. The closest relative of *A. patagonica* is *Archiphora robusta*, in Baltic amber (Hennig, 1964; McAlpine and Martin, 1966), again a highly derived position (fig. 6.5). *Archiphora* undoubtedly had a much more widespread distribution than it does today, and it is certainly relict. The family positions of *Sciadophora* and *Prioriphora*, in Canadian Cretaceous amber, have been debated. Both have been placed in the Sciadoceridae by McAlpine and Martin (1966), but I have indicated that *Prioriphora* is a plesiomorphic phorid (Grimaldi 1989). The important point is that, with knowledge of the Baltic amber *Archiphora*, it would be naive to propose an austral landmass vicariance as the event solely responsible for present-day sciadocerids. Proposing an austral vicariance for the living sciadocerids, in lieu of *Archiphora robusta*, would be proposing a hypothesis based on a paraphyletic group.

Cretaceogaster is another instance of Diptera fossilized in the Canadian Cretaceous amber. The fossil genus *Cretaceogaster* is the sister group to the extant Chilean genus *Parhadrestia* (fig. 6.6). Both genera represent the sister group to the rest of the stratiomyids (soldier flies), a speciose cosmopolitan group (Woodley 1986). Since, by outgroup comparison, the primitive state for the distribution of the *Cretaceogaster* + *Parhadrestia* clade is cosmopolitan, and the distribution of this clade obviously lies within the bounds of a cosmopolitan distribution, one cannot conclude that the distribution of the

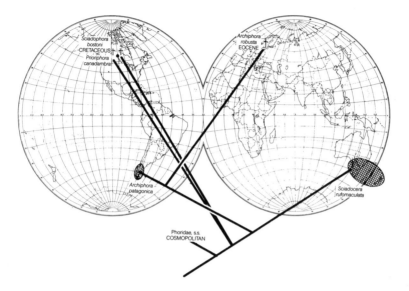

Fig. 6.5. Distribution of living and fossil Sciadoceridae (Diptera) and relatives. (From Hennig 1964; McAlpine and Martin 1966).

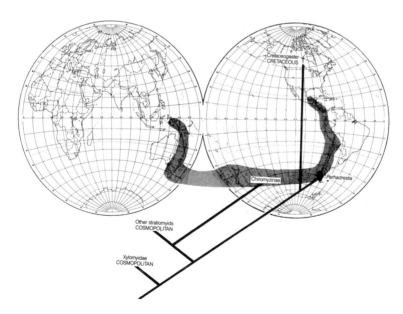

Fig. 6.6. Distribution of primitive living Stratiomyidae (Diptera) and a Cretaceous amber fossil. (From Woodley 1986.)

clade necessarily was any larger than is presently known. Any number of fossil sister groups to the *Cretaceogaster + Parhadrestia* clade could be discovered anywhere in the world, which would not affect the conclusion that the family is cosmopolitan. However, the amber fossil does put a minimum age on the *Parhadrestia* clade, thus allowing one to narrow the choice of geological events potentially influential in stratiomyid distributions.

North American Fossil Tsetses

Perhaps no insect fossils from the Oligocene shales of Florissant, Colorado, are better known than the tsetse flies. They are famous as an obvious example of geographic extinction, for today *Glossina* occurs only in central Africa and the southern tip of the Arabian peninsula (a living example is shown in figure 6.7). These fossil flies have inspired considerable speculation, particularly with regard to their possible effects on North American Tertiary mammals. Apparently, Henry Fairfield Osborn, the sometimes bombastic vertebrate paleontologist from the American Museum of Natural History, hypothesized that the rampant extinction of many large mammals from the North American Tertiary was due to disease, perhaps transmitted by biting flies. Several years later, when T. D. A. Cockerell, the famous insect paleontologist, described several fossil tsetses, he became enamored by Osborn's idea (Cockerell 1907, 1908, 1918). Despite criticism of the hypothesis on the grounds that native mammal hosts of living *Glossina* are immune to trypanosomiasis, Cockerell (1919) still defended the view. Even recently, fossil tsetses have been involved in a scenario speculating on extinction of australopithecine hominids and the success of *Homo* (Lambrecht 1985). The systematics and phylogenetic position of the Florissant specimens has direct bearing on whether these fossil flies carried trypanosomes and whether the modern African tsetse flies have become restricted in range.

Reexamination of eight of the nine known fossil glossinids shows that two species exist, not four (see appendix). An important synapomorphy of living *Glossina* is missing in the fossil species, even though much finer detail is preserved. This indicates that the fossil species are the sister group(s) to the monophyletic living species (see appendix; fig. 6.8). To state that *Glossina* had a formerly more widespread distribution is to use a broader definition of the genus than presently applies to modern species (i.e., a diagnosis omitting the characters of arista, legs, wing membrane, and perhaps male genitalia). In effect, one is extending the modern distribution by adding an outgroup and broadening the diagnosis of the modern group, which still does not affect or

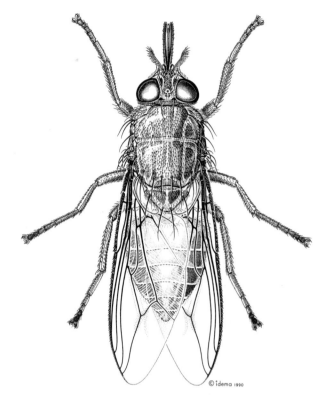

Fig. 6.7. Dorsal habitus of *Glossina pallidipes* female (Africa, recent).

explain historical biogeography of living *Glossina* in Africa. Many features support the fact that the sister group to the Glossinidae is the Hippoboscidae *sensu lato* (Pupipara in Griffith's sense) (Hennig 1973; Griffiths 1972; 1976). McAlpine (1989) maintained that a sister-group relationship of the Glossinidae is only to the Hippoboscidae (s.s.) and that these two families together are the sister group to the bat flies (the Steblidae and Nycteribiidae). The Hippoboscidae, in any sense, are cosmopolitan, so outgroup comparison alone does not suggest a formerly wider distribution of living *Glossina*. However, were the Hippoboscidae restricted to Africa, a contiguous distribution of North American Oligocene and living African Glossinidae would be more compelling. There are no African glossinid fossils. The discovery that the Florissant specimens are the sister group(s) of the African *Glossina* invites similar cladistic analyses with various other taxa showing North American

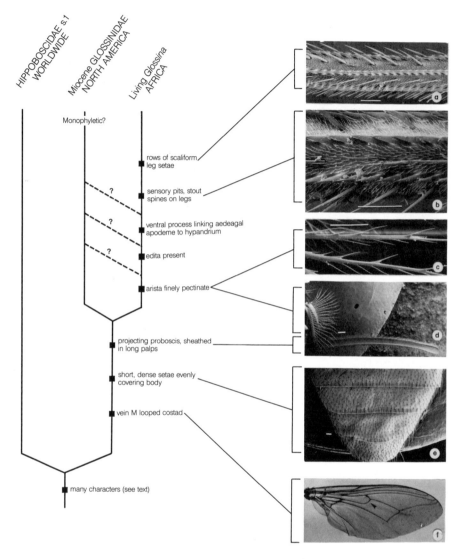

Fig. 6.8. Cladogram of living and extinct Glossinidae (a–f). Scanning electron micrographs of taxonomically important features of living *Glossina*.

Tertiary–modern African connections. These include numerous mammals from as recent as the Pleistocene (e.g., *Acinonyx* [cheetah], *Chastaporthetes* [hyena], *Camelops* [camel], *Mammut* and *Mammathus* [Elephantidae]—see Kurtén and Anderson [1980], Anderson [1984]).

There is even at least one other Florissant, Oligocene insect group with a distribution very similar to that of the Glossinidae, namely, the spoon-winged lacewings, family Nemopteridae (Neuroptera). *Marquettia americana* (Cockerell) is known from the Florissant beds, and another (but very poorly preserved) specimen of a nemopterid is known from Oligocene shales of Montana. Nemopteridae presently occur in southern Europe, the Middle East, India, Africa, and Chile, and are absent from North America. Even though the classification of the Nemopteridae presently rests upon characters of wing venation (which is exquisitely preserved in the Florissant specimen), there has not been any phylogenetic study incorporating *Marquettia* into a revision of living nemopterids. Carpenter, who has studied both North American fossil specimens, stated that *Marquettia* is related "to several stenonemiine genera, but it stands apart from all of these by the extensive development of the radial sector. In this respect, I consider the genus *Marquettia* to be the most primitive of the genera of the Nemopteridae known at the present time" (1959:21). If *Marquettia* is the sister group to the living Nemopteridae, this would reflect the situation found in the Glossinidae.

Vicariance hypotheses are obviously not fossil dependent, nor can fossils necessarily refute vicariant hypotheses. As shown here, fossils can force the revision of a vicariance hypothesis by extending an area of endemism based only on living taxa. This is not a feature limited to fossils, and it requires restricted conditions. Certainly the austral distributions of living midges, sciadocerids, archaeid spiders, and many other arthropods (e.g., see Hennig 1965) present irrefutable evidence of a pervasive geological event in the southern land masses, such as the breakup of Gondwanaland. Austral landmass drift occurred within the Cretaceous period (the Baltic amber, for example, is Eocene to Miocene). The objection can be raised that Baltic extinctions represent younger, extinct dispersal elements from the south and therefore are unimportant in understanding southern patterns. However, the Baltic amber taxa may quite easily have been contemporaneous with Cretaceous austral taxa (the Eocene is a minimum age for some of the amber and its inclusions). Moreover, the number of austral taxa with extinct relatives in Baltic amber is as much a pattern to deal with as the patterns of living austral taxa, for which the reader is referred to Ander's (1942) discussion of many other examples.

An extreme argument on the role of fossil data in biogeography is that of Eskov (1987). He disagrees with the vicariant biogeographic pattern discussed earlier by Forster and Platnick (1984), mostly on his use of poorly preserved spider fossil impressions from the Jurassic of the Soviet Union. He argued that austral or gondwanan distributions are entirely reflective of extinction, best understood not by phylogenetics, but by fossils. Unfortunately, virtually all his examples lack any phylogenetic analyses. For many of his examples the fossils either are too poorly preserved to allow accurate placement in a scheme of living relatives or may be the sister group to living relatives. The claim that "southern continents have indeed inherited a lot of relict taxa extinct from the north" (1984:92) makes numerous presumptions. The approach of Eskov (1987:102) is entirely the reverse of what is espoused here: a Southern Hemisphere disjunction must be regarded as a result of pervasive tectonics until the alternative (a pancontinental range) can be shown to be a reasonable alternative.

Analyses of the sort discussed here depend fundamentally on being able to survey in a fossil all the characters required to place it accurately in a phylogeny of living relatives. One will note that all my examples, except for the exceptional case of the tsetses, are of amber fossils; this is because amber fossilizes insect details much better than any type of mineral replacement fossilization. Systematists working on extant species are more inclined to include amber fossils in their studies than to include mineral replacement specimens. Numerous examples of geographic extinctions from the amber fossil record are given in Larsson (1978) and especially Ander (1942), to which I refer the reader for particularly stimulating accounts.

Appendix:
Systematic Paleoentomology

The species status of all nine known fossil specimens, apparently belonging to four species, has been untested for nearly seventy years since Cockerell described the last of his tsetse fossils (e.g., Bequaert 1930). A detailed comparative study has never been made of all the specimens; Cockerell, in fact, had described new species and deposited them in various museums as each specimen had been discovered. Cockerell (1907) gave an account of the fossil insect-collecting localities at Florissant, as well as a general account (1937). A comprehensive review of the flora and fauna of the Florissant strata is provided by MacGinite (1953); Axelrod (1987) monographed a slightly younger flora in Colorado, which provides interesting comparisons.

Very few diagnostic features are available on the Florissant tsetse fly specimens, despite the beautifully preserved detail. Those that *are* preserved and that are useful at the generic and species level are venation, fine setal vestiture on integument, spination of legs, chaetotaxy (in some specimens) of thorax and abdomen, and length of (at least a portion of) the mouthparts. Male genitalia, details of the chaetotaxy of the thorax and head, and mouthpart structure are some of the features required in delimiting species in modern material, but they are not present in the fossils. Thus emphasis was placed on relative lengths in wing venation in sorting the Florissant species.

Table 6.1. Ratios of Wing Measurements for fossil *Glossina.**

Specimen		Ratios			
		A	B	C	D
MCZ 3490		0.53	1.21	0.47	0.21
AMNH 18839	l	0.56	1.24	0.49	0.18
	r	0.51	1.11	0.48	0.21
BMNH 8421	l	0.52	—	—	0.24
	r	0.55	1.03	0.54	0.21
AMNH 295		0.56	1.17	—	0.20
BMNH 19223		0.36	0.66	0.38	0.23
NMNH 66282		0.40	0.63	0.30	0.21
NMNH 66281	l	—	1.32	0.54	0.23
	r	0.47	1.23	—	0.20
AMNH 39554		0.47	1.14	0.60	0.22

* See text for definitions of ratios.

Among the venational characters used, the following were five measurements and four ratios that were made.

1. Length of vein R_{4+5} along cell *br*, beginning at junction of vein R_{2+3} and R_{4+5}.
2. Greatest width of cell *br*.
3. Length of vein M_{1+2} from end of vein *rm* to junction of vein M.
4. Distance between end of vein Sc and end of cell *br*.
5. Length of vein R_{4+5} from vein *rm* to apex.

For the ratios: (A) 1/5, (B) 1/3, (C) 4/1, and (D) 2/1.

Drawings of the specimens were made with the use of a camera lucida, with care being taken to avoid distortion, particularly around the periphery of the camera lucida image. Measurements were made directly from the drawings, and the measurements were not converted to millimeters, but ratios were made directly from these. Table 6.1 gives the four wing vein ratios for eight of the nine known specimens; despite attempts to borrow the ninth specimen, supposedly very complete and well preserved, there was no compliance by the University of Colorado Museum. Ratio D (width of the characteristically shaped cell *br* relative to a portion of its length), was the only ratio found not diagnostic of the species limits described later. Figure 6.9 shows a bivariate scatter plot of ratios A and B. For some ratios where there were measurements from the right and left halves, the means of the two values were put on the plot.

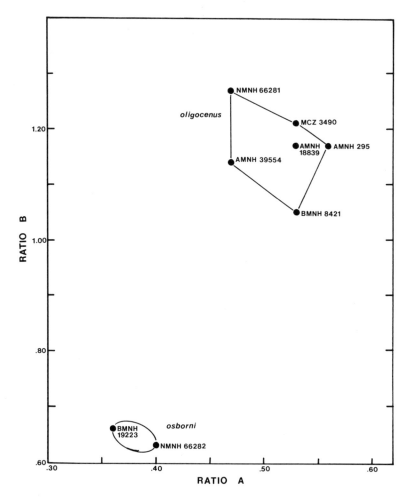

Fig. 6.9. Bivariate plot of wing ratios in fossil glossinids. (See text for discussion of ratios.)

Glossina oligocenus (Scudder)

Paloestrus oligocenus Scudder, 1892: 19. Type: MCZ 3490 a, b, from Oligocene beds at Florissant, Colorado.

Glossina oligocenus: Cockerell, 1907: 414 (for AMNH 18839?)

Glossina oligocena (sic): Cockerell, 1908: 65 (for MCZ 3490, and another uncatalogued specimen, perhaps AMNH 18839).

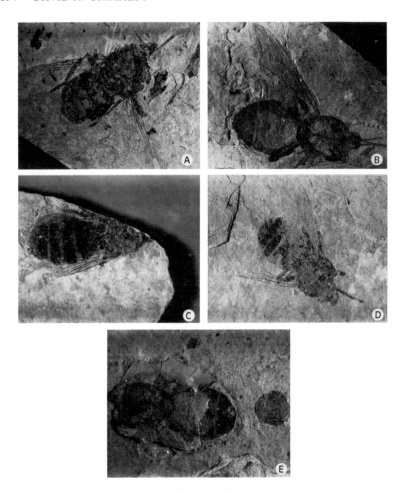

Fig. 6.10.A–E Habitus photos of fossil glossinids: (A) AMNH 18839, (B) AMNH 295, (C) NMNH 66282, (D) NMNH 66281, (E) AMNH 39554.

Glossina veterna (Cockerell, 1916: 70. Type: NMNH 66281; Cockerell, 1918: 310, pl. 55 (general account). NEW SYNONYM.

Diagnosis. Very large glossinid, wing length 14.0–18.0 mm, with characteristic shape of upcurved *br* wing cell typical of modern species; distance between levels of end of subcostal vein and distal end of cell *br* at least one-third the length of cell *br*.

Material Examined. Types indicated earlier, and AMNH 39554, AMNH 295, BMNH 8421, AMNH 18839.

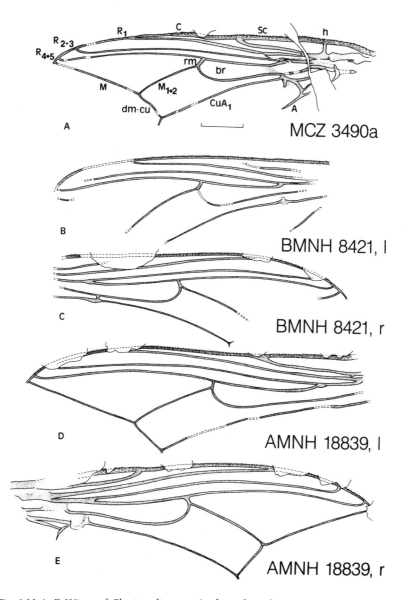

Fig. 6.11.A–E Wings of *Glossina oligocenus* (scale = 2 mm).

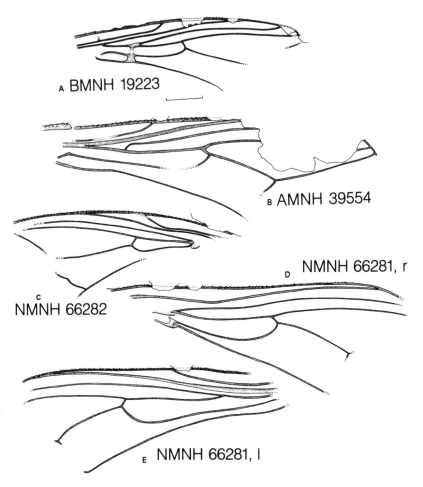

Fig. 6.12.A–E Wings of *Glossina osborni* (as newly revised) (A, C) and *G. oligocenus* (as newly revised) (B, D, E) (scale = 1 mm).

Description of Specimens.

MCZ 3490 (reverse and obverse impressions). This is a nearly complete specimen, with wings and some legs nearly completely preserved. Lack of proboscis apparently had caused Scudder to believe the specimen to be an oestrid. Wing length about 15.5 mm. Relief of the basal portion of wing is apparent, helping to identify veins in this area. Five small pegs occur on costal vein of obverse (a) specimen,

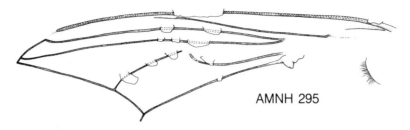

AMNH 295

Fig. 6.13. Wing of *Glossina oligocenus*, AMNH 295.

where subcostal vein meets C and before R_1 and C meet. The wing of this specimen is the most complete of all the fossil specimens.

AMNH 39554. This is identified by Cockerell as G. *veterna*, based on label with specimen in his handwriting but never reported in the literature by him. This is a unique specimen, one aspect concerning the detail preserved at the apex of the abdomen. Here the ventral side is exposed, showing two paramedian lobes with numerous fine pits (probably setal sockets), the lobes of which are probably the cerci. There are also two median lobes, one just anterior, the other just posterior to the cerci. Also, the origin of vein R_{2+3} from the base of R_{4+5} is nearly perpendicular, not extremely acute, as in all other specimens of *oligocenus*. This may be just an artifact or an extreme variant, as all other venational features compare well with other specimens of *oligocenus*. Someone had apparently attempted patching the specimen but destroyed much of it with glue.

NMNH 66281 (obverse and reverse). This is a very complete, well-preserved specimen, the body of which shows the dorsal aspect. Both wings are preserved and splayed away from the body; the head and proboscis, thorax, and abdomen are intact. The tergal and scutellar bristles, typical of living *Glossina,* are preserved. Cockerell (1916) described *veterna* on the basis of intermediate size (between *oligocenus* and *osborni*) and lower margin of wing cell *br* "not bulg[ing] much near the end." The shape of cell *br* is actually indistinguishable from other specimens of *oligocenus*. Cockerell's measurement of wing length (10.9 mm) must be a minimum estimate, as much of the apex and base of both wings are not intact.

BMNH 8421. Specimen has dorsal surface exposed. Portion of one side of the head is present; the proboscis is lost. Most of the thorax and abdomen is intact, and the fine setulae covering the abdomen are very well preserved. Portions of four legs are preserved, but it is impossible to determine which pairs they are.

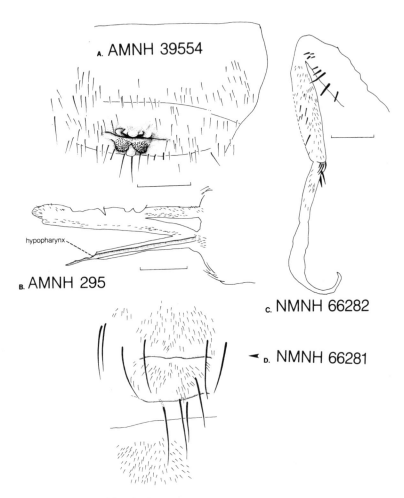

Fig. 6.14.A–D Parts of fossil *Glossina*: (a) terminalia, ventral view, AMNH 39554, (b) proboscis, AMNH 295, (c) leg, NMNH 66282, (d) scutellum and posterior margin of notum, NMNH 66281.

AMNH 295, 18339. Both of these specimens were presumably unreported in the literature by Cockerell. 18839 possesses a portion of both wings, with most veins intact. A (hind?) leg shows a row of 10 short pegs on the first tarsomere; setae on the proboscis and its base are preserved, as well as a channel along the middle of the base of the proboscis. AMNH 295 is nearly complete, with right wing, abdomen, thorax, head and proboscis, and portions of (hind?) legs

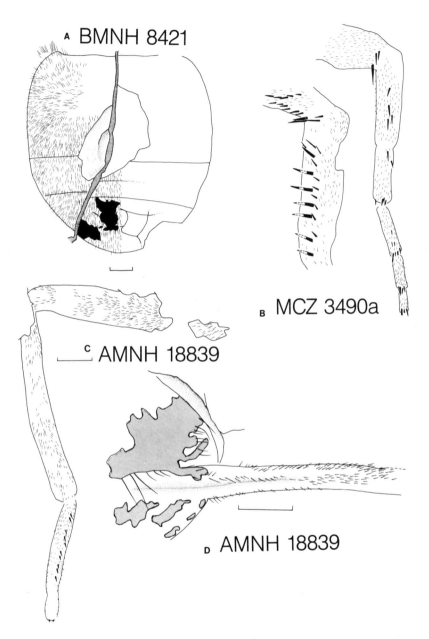

A BMNH 8421

B MCZ 3490a

C AMNH 18839

D AMNH 18839

Fig. 6.15.A–D Parts of *Glossina oligocenus* (as newly revised): (A) entire abdomen, ventral view, (B) Legs, (C) Legs, (D) base of proboscis and anterior margin of head.

preserved. The hypopharynx of this specimen is separated from the rest of the proboscis. The specimen was apparently collected on the 1906 expedition to Colorado.

Glossina osborni Cockerell

Glossina osborni Cockerell, 1909: 128. Type: BMNH 19223.

Glossina armatipes Cockerell, 1917: 19. Type NMNH 66282; Cockerell, 1918: 310–311 (review). NEW SYNONYM.

Cockerellitha Townsend, 1938: 166. Monotypic, Type: *osborni*.

Lithoglossina Townsend, 1938: 166. Monotypic, Type *armatipes*.

Glossina osborni; Hegh, 1946: 102. (Synonymy of Townsend's genera).

Glossina osborni; Hennig, 1973: 337 (Synonymy of Townsend's genera).

Diagnosis. Smaller glossinid, wing length approximately 7–11 mm.; wing cell *br* upcurved, relatively more slender than in *oligocenus*; distance between levels of end of subcostal vein and distal end of cell *br* barely longer than length of vein *rm*; cell *br* not extended to base of wing, closed approximately midway between apex of *br* and wing base.

Material Examined. Two type specimens indicated earlier.

Description of Specimens.

BMNH 19223 (lateral view preserved, with one wing). All the thorax and abdomen is preserved, but only the basal portion of the head and several legs are preserved (which pairs is not evident). No trace of setae on thorax or abdomen; dense vestiture of fine, short setulae covering entire abdomen (best seen under layer of ethanol). Proboscis not preserved (Cockerell [1906] mentions "proboscis evident, though imperfect"—he may have been referring to one leg). Wing length about 7 mm. Cockerell mentioned that "the vein bounding the outer side of the discal [br] cell has a double curve." Presumably he was referring to vein *rm*, which, in *osborni*, is not slanted proximad as much (or even slightly concave) as it is in *oligocenus*.

NMNH 66282. Cockerell (1917) gave a very complete description of this specimen. Wing length about 7.5 mm, according to Cockerell; the base and apex of the wing are very incomplete, but total length appears to be longer than 7.5 mm. Apparently, *hind* femur bearing a row of eight long bristles, according to Cockerell; the only way in which this leg was discerned as being a hind one is that the living species *G. fusca* possesses similar spines on the hind femur. These are probably the same bristles observed in MCZ 3490, which are specimens of *G. oligocenus*. Thus, femoral bristles would not be a feature distinctive to NMNH 66282. Apparently, femoral bristles were not preserved in many of the specimens, either because of the fact that legs have been lost, or preserved portions are not in suitable orientations, or because of the fact that large bristles are very easily dislodged from an insect specimen. Another feature that Cockerell used to diagnose *G. armatipes* (NMNH 66282) is the position of crossvein *rm* relative to vein R_{4+5}. This crossvein is more oblique in NMNH 66282 than in BMNH 19223, but only slightly more so; neither of these two specimens has vein *rm* lying nearly as oblique to R_{4+5} as occurs in most specimens of *oligocenus*. In fact, the angle of *rm* is quite variable, as seen in the nearly perpendicular *rm* vein in AMNH 18339 (*oligocenus*).

Discussion of Glossinid Systematics

A composite of the Florissant specimens, including both species as newly revised (*oligocenus* and *osborni*), would reveal the following features: wing with a medial + crossvein *r-m* looped costad; microtrichia on wing arranged in regular, longitudinal rows; body (particularly the abdomen and including legs) stout, densely covered with fine, short setulae; proboscis projecting, ensheathed by long palps; both species presumably with long, stout spines on hind (?) femora. As is characteristic of living material, modern *Glossina* species show several other morphologic features, which indicate they are a monophyletic group: for male genitalia (being internal, which would hardly be preserved in the fossils), aedeagal apodeme linked to the hypandrium by a ventral process about near its middle, and "edita" (paramere) present (Griffiths 1972). Also, in modern *Glossina* the arista has a very distinctive array of primary, secondary, and tertiary branches; wing membrane has a fine, random, corrugated pleating; legs have rows of sensory pits, stout spines, and scaliform setae. Neither the aristal, the wing membrane pleating, nor leg characters are present in the Florissant fossils, and presumably they should have been preserved, given that even the fine setulae of the abdomen and legs and (in some specimens) the wing membrane microtrichia have been preserved. Thus, one can at most hypothesize a sister-group relationship of the Oligocene *Glossina* to modern, Afrotropical ones, for the Oligocene specimens are apparently plesiomorphic to the living *Glossina*. I would not propose a new genus or even a subgenus for the two Oligocene species, as there is no evidence that they together represent a monophyletic group.

ACKNOWLEDGMENTS

I am grateful to the curators who loaned me fossil *Glossina* specimens in their care: Frank Carpenter (Museum of Comparative Zoology, Harvard University), Paul Whalley (British Museum), and Frederick Collier (Paleobiology, Smithsonian Institution). I am grateful too for the input on the manuscript given by Norman Platnick, Norman Woodley, Ross MacPhee, Mike Novacek, and Quentin Wheeler.

REFERENCES

Ander, K. 1942. Die insektenfauna des Baltischen bernsteins nebst damit verknüpften zoogeographischen problem. *Lunds Univ. Årsskrift* 38:3–82.

Anderson, E. 1984. Who's Who in the Pleistocene: A mammalian bestiary. In P. S. Martin and R. G. Klein, eds., *Quaternary Extinctions: A Prehistoric Revolution.* University of Arizona Press, 892 pp.

Axelrod, D. I. 1987. The late Oligocene Creede Flora, Colorado. Univ. California Publ. Geol. Sci. 130: x + 166 pp., 34 plates.

Bequaert, J. 1930. Tsetse flies—past and present (Diptera: Muscoidea). *Ent. News* 41:158–164, 202–203,227–233.

Brundin, L. 1966. Transantarctic relationships and their significance, as evidenced by chironomid midges, with a monograph of the subfamilies Podonominae and Aphroteniinae and the Austral Heptagyiae. *Kungl.* Svenska Vet. Hand. 11:3–472, 30 plates.

Brundin, L. 1981. Croizat's panbiogeography versus phylogenetic biogeography. In G. Nelson and D. E. Rosen, *Vicariance Biogeography*: A Critique, pp. 94–138. New York: Columbia University Press.

Brundin, L. Z. 1990. Phylogenetic biogeography. In A. A. Myers and P. S. Giller, *Analytical Biogeography.* London: Chapman and Hall.

Carpenter, F. M. 1959. Fossil Nemopteridae (Neuroptera). *Psyche* 66:20–24.

Cockerell, T. D. A. 1907. An enumeration of the localities in the Florissant Basin, from which fossils were obtained in 1906. *Bull. Amer. Mus. Nat. Hist.* 23(4):127–132.

Cockerell, T. D. A. 1907. A fossil tsetse-fly in Colorado. *Nature* 76:414.

Cockerell, T. D. A. 1908. Fossil insects from Florissant, Colorado. *Bull. Amer. Mus. Nat. Hist.* 24:59–69.

Cockerell, T. D. A. 1909. Another fossil tsetse fly. *Nature* 80:128.

Cockerell, T. D. A. 1916. A third fossil tsetse-fly. *Nature* 98:70.

Cockerell, T. D. A. 1917. A fossil tsetse fly and other Diptera from Florissant, Colorado. *Proc. Biol. Soc. Wash.* 30:19–22.

Cockerell, T. D. A. 1918. New species of North American fossil beetles, cockroaches, and tsetse flies. Proc. U.S. Nat. Mus. 54:301–311, plates 54–55.

Cockerell, T. D. A. 1919. *Glossina* and the extinction of Tertiary mammals. *Nature* 103:265.

Cockerell, T. D. A. 1937. Recollections of a naturalist, v. fossil insects. *Bios* 8:51–56.

Eskov, K. 1987. A new archaeid spider (Chelicerata: Araneae) from the Jurassic of Kazakhstan, with notes on the so-called "Gondwanan" ranges of recent taxa. *N. Jb. Geol. Paläont. Abh.* 175:81–106.

Forster, R. R. and N. I. Platnick. 1984. A review of the Archaeid spiders and their relatives, with notes on the limits of the superfamily Palpimanoidea (Arachnida, Araneae). *Bull. Amer. Mus. Nat. Hist.* 178:1–106.

Grande, L. 1985. The use of paleontology in systematics and biogeography, and a time control refinement for historical biogeography. *Paleobiology* 11:234–243.

Griffiths, G. C. D. 1972. The phylogenetic classification of the Diptera Cyclorrhapha, with special reference to the male postabdomen. The Hague: W. Junk.

Griffiths, G. C. D. 1976. Comments on some recent studies of tsetse-fly phylogeny and structure. *Systematic Entomology* 1:15–18.

Grimaldi, D. 1989. The genus *Metopina* (Diptera: Phoridae) from Cretaceous and Tertiary ambers. *J. N.Y. Entomol. Soc.* 97:65–72.

Hegh, E. 1946. Les Tsé-tsés: description, biologie—moyens de destruction. Brussels: Royaume de Belgique. 102 pp.

Hennig, W. 1964. Die dipteren-familie Sciadoceridae im Baltischen bernstein (Diptera: Cyclorrhapha Aschiza). *Stutt. Beitr. Naturk.* 127:1–10.

Hennig, W. 1965. Die Acalyptratae des Baltischen bernsteins. *Stutt. Beitr. Naturk.* 145:1–215.

Hennig, W. 1973. Diptera. In *Kükenthal's Handbuch der Zoologie* 4(2) 31:1–337. Berlin: Walter de Gruyter.

Humphries, C. and L. Parenti. 1986. *Cladistic Biogeography.* New York: Oxford University Press.

Kurtén, B., and E. Anderson. 1980. Pleistocene mammals of North America. New York: Columbia University Press.

Lambrecht, F. L. 1985. Trypanosomes and hominid evolution. *Bioscience* 35:640–646.

Larsson, S. G. 1978. Baltic Amber—A Paleobiological Study. Klampenborg: Scandinavian Science Press.

McAlpine, J. F. 1989. Phylogeny and classification of the Muscomorpha. *In* J. F. McAlpine, ed., *Manual of Nearctic Diptera.* Research Branch Agriculture Canada Monograph 32, 3:1397–1518.

McAlpine, J. F. and J. E. H. Martin. 1966. Systematics of Sciadoceridae and relatives with descriptions of two new genera and species from Canadian amber and erection of family Ironomyiidae (Diptera: Phoroidea). *Can. Entomol.* 98:527–544.

MacGinite, H. D. 1953. Fossil plants of the Florissant Beds, Colorado. Carnegie Instit. Wash. Publ., Contr. Paleont. 599:1–98, 75 plates.

Marshall, L. G. 1990. Extinction. In A. A. Myers and P. S. Giller, eds. *Analytical Biogeography.* London: Chapman and Hall.

Michener, C. D. 1982. A new interpretation of fossil social bees from the Dominican Republic. *Sociobiology* 7:37–45.

Michener, C. D. and D. A. Grimaldi. 1988. A *Trigona* from Late Cretaceous amber of New Jersey (Hymenoptera: Apidae: Meliponinae). *Amer. Mus. Novit.* 2917:1–10.

Patterson, C. 1981. Methods of paleobiogeography. In G. Nelson and D. E. Rosen, eds., *Vicariance Biogeography: A Critique,* pp. 446–489. New York: Columbia University Press.

Platnick, N. I. and G. Nelson. 1978. A method of analysis for historical biogeography. *Syst. Zool.* 27:1–16.

Scudder, S. H. 1892. Some insects of special interest from Florissant, Colorado and other points in the territories of Colorado and Utah. *Bull. U.S. Geol. Surv.* 93: 25 pp., 3 pls.

Townsend, C. H. T. 1938. Five new genera of fossil Oestromuscaria (Diptera). *Entomol. News* 49:166–167.

Vermeij, G. 1989. Geographical restriction as a guide to the causes of extinction: the case of the cold northern oceans during the Neogene. *Paleobiology* 15:335–356.

Wille, A. 1977. A general review of the fossil stingless bees. *Rev. Biol. Trop.* 25:43–46.

Woodley, N. E. 1986. Parhadrestiinae, a new subfamily for *Parhadrestia* James and *Cretaceogaster* Teskey (Diptera: Stratiomyidae). *Syst. Entomol.* 11:377–387.

7 : Extinction, Sampling, and Molecular Phylogenetics

Ward C. Wheeler

Abstract. Extinction, whether natural or artificial, removes taxa from phylogenetic consideration. Here the effects of extinction on molecular data are examined through computer simulations. An analysis of variance (ANOVA) analysis is performed to examine the effects of several variables on the accuracy and resolution of phylogenetic reconstruction. Overall, the number of taxa used in cladogram construction is the most important factor in cladogram accuracy and the number of characters is most responsible for its resolution.

Extinction poses a dilemma for all systematics; molecular studies are no exception. For molecular systematists, the question is fundamental: Do taxa lost through extinction affect our interpretation of the relationships among living organisms? Using morphology, Gardiner (1982), following Patterson (1981), assumed in his study of amniotes that they do not. Gardiner constructed a phylogeny based solely on extant taxa, erecting the group Homeothermia, which unites birds and mammals. However, Gauthier et al. (1988) demonstrated persuasively that the inclusion of fossil taxa can greatly affect our phylogenetic understanding of these same organisms. In their analysis, the inclusion of fossil synapsids caused the position of mammals to shift relative to lepidosaurs and chelonians (turtles). The influence of fossil taxa in such realignments was further emphasized by Donoghue et al. (1989). With very few exceptions, fossil organisms cannot be employed in molecular studies. In the construction of phylogenies can we compensate for this deficit in fossil taxa by turning instead to the wealth of molecular characters available from extant organisms?

The limitations imposed by natural extinction on molecular studies is further compounded by a form of artificial extinction, in which systematists draw on characters from a single taxon to represent a large and diverse group (Wheeler 1989). In these cases, the characters of that exemplar are assumed to reflect the ground plan of the entire group, but the establishment of this ground plan is impossible, throwing doubt on the synapomorphies linking higher taxa.

Analysis of phylogenetic relationships can also be hampered by missing data. In morphologic studies, the poor quality of a specimen or a condition of extreme apomorphy may render characters unobservable. In molecular systematics, on the other hand, the type of missing data encountered most often is a gap or gaps in aligned sequences. As with the loss of taxa through extinction, molecular systematists propose that the problem of missing data can be overcome by gathering more data.

Both missing characters and extinct taxa dilute the pool of data available to investigate phylogenetic patterns. In the first case, characters are lost, and in the second, taxa are lost, but they amount to the same thing: a decrease in the sample of phylogenetic information. Computer simulation is used here to examine the effects of missing characters and taxa on the efficacy of phylogenetic reconstruction based on molecular sequence data.

Simulations

There were three parts to the simulation procedure: the generation of the sequences, the culling of taxa, and the analysis of the phylogenetic results.

Evolution with both constant and variable rates of change was simulated. In the first model, the rate of anagenetic change was held constant throughout all lineages and for all characters. Similarly, the rate of cladogenesis was kept constant for all lineages. The second evolutionary model was more complex. Here the rates of evolution were determined uniquely for each character by a Poisson process. The probability of each lineage splitting was determined by another Poisson process. Thus, although the constant model simulated a clocklike evolution of characters and lineages, the variable model depicted a system with changing rates of character evolution and lineage splitting.

These two models of evolution were simulated using 25 different ratios of anagenesis versus cladogenesis. These ratios ranged from a splitting probability 50 times that of character change (hence a character has a 1 in 50 chance of transforming before the next lineage split) to a splitting probability one third that of character change. This rate variation yields scenarios wherein there is little (if any) character evolution between speciation events to those in which a great deal of evolution has occurred in the interval between lineage splits. At the most rapid rates of character change, the sequences approached randomness due to the extreme amounts of character change.

In both models, the nucleotide distribution was entirely symmetrical ($A = C = G = T = 0.25$), as were the transformation probabilities among bases (all transformations equally likely). Sequences of three lengths—50, 100, and 250 bases—were generated.

All these permutations were repeated for varying proportions of missing data. There were four levels of deficit 0%, 5%, 10%, and 25%. Two procedures for choosing characters and taxa were employed. In the first, all characters were equally likely to become missing, whereas in the second, a Poisson distribution was used to establish a character's probability of becoming missing. All taxa had an equal chance of losing characters in both cases.

The purpose of the Poisson distribution was to add more variation to the system ($\mu = \sigma^2$, for Poisson). Under its influence, certain lineages and characters will evolve more rapidly than others. Similarly, certain base positions were more likely to be unobservable or missing. Both these situations seemed to represent the circumstances of evolution more realistically than simple homogeneous change due to the mechanical shortcomings of sequencing protocols, alignment procedures, and observed phenomena such as mutational "hot spots."

The second part of the simulation concerned the extinction process. Twelve sequences were generated for each data set using the preceding procedures, but only those with four monophyletic groups (with two or more taxa each) were retained. One representative sequence was chosen from each of these four groups at random, and a new, less diverse data set was created. This culling process was repeated 25 times for each of the data sets, to produce one full data set and 25 altered ones. Thus 26 data sets were generated for each combination of parameters: model of evolution, percent missing data, type of missing data, length of sequence, and ratio of anagenesis to cladogenesis.

These data sets were in turn subjected to Farris' HENNIG 86 (1988, Version 1.5) to determine the most parsimonious phylogenetic arrangement of the taxa. In these runs, the characters were considered nonadditive (no transformation series among states was specified, hence no specific evolutionary model was assumed). The heuristic options mh★bb★, the most exhaustive of the heuristic tree searches involving the construction of multiple trees and branch-swapping, were used because implicit enumeration proved too time-consuming. When multiple, equally parsimonious solutions existed, the strict consensus method was used to summarize agreement among the most-favored hypotheses.

The results of these phylogenetic analyses were then tested for their similarity to the "correct" or known tree, as reported from the sequence-generating program. In the case of the four-taxa data sets, the tree produced either agreed completely with the arrangement of the four monophyletic groups in the original data or did not agree at all. The situation was more complex for the full, 12-taxa data set. Here concordance between the known tree and that produced was a matter of degree, with some members of the monophyletic groups correctly placed and some not. Thus agreement was assessed as the product of a series of fractions, representing the proportion of correctly placed members in each of the four monophyletic groups.

In the analysis, a further distinction is made between arrangements of taxa that are correctly placed and those that are not incorrectly placed or unresolved. An unresolved cladogram may not be "correct"—it neither conveys crucial information nor is necessarily misleading. Polytomous hypotheses make no statements about certain potential groups and avoid the "incorrect" placement of taxa. A fully resolved tree, on the other hand, can be utterly disinformative.

The level of resolution of the 12-taxa cladograms was defined as the number of resolved groups divided by the number of potentially resolved groups. For 12 taxa there are 10 nontrivial groups (all 12 always form a group); hence the index of resolution is the number of resolved groups divided by 10.

Results

Five factors affected the outcome of the phylogenetic trials: the number of taxa (12 or four), the length of the sequence modeled (number of characters), the type and degree of missing data, the rate of evolution, and the stochastic-model under which the sequences "evolved." All five conditions determined the percent of reconstructions that were not incorrect. Four of the factors were used to assess cladogram resolution; the number of taxa was not included in this assessment since the level of resolution only had meaning for the full data sets.

The results of the trials were evaluated through two multiway analyses of variance (ANOVA) procedures. In the first of these, each of the five factors was treated as an independent variable, with the success of the cladogram (percent not incorrect) as the dependent variable. The second analysis was a four-way procedure with the level of cladogram resolution as the dependent variable. Both of these ANOVAs were models without replication. Therefore interaction effects could not be separated from residual error. The ANOVA results are summarized in tables 7.1 and 7.2. Since the dependent values were percents, the data transformation $\arcsin \sqrt{\theta}$ was used to check normality. The additivity assumption of this model was examined with Tukey's test for additivity. For both ANOVA procedures—with and without data transformation—there was insignificant deviation from additivity (table 7.3).

As these tables show, four of the five factors were significant contributors to variation in cladogram success, as measured by fraction not incorrect: the

Table 7.1. Factors Affecting Cladogram Success (Percent not Incorrect)

Factor	df	Unaltered Data		Transformed Data	
		F	% variance	F	% variance
Number of taxa	1	11.9*	47.3	13.2	49.7
Sequence length	2	4.52†	17.9	4.57†	17.2
Missing data	6	0.22	0.89	0.28	1.07
Rate of evolution	24	2.70*	10.7	2.72*	10.2
Stochastic model	1	4.84†	19.9	4.77†	18.0

* Significant at $p < 0.001$.
† Significant at $p < 0.05$.

Table 7.2. Factors Affecting Cladogram Resolution

Factor	df	Unaltered Data		Transformed Data	
		F	% variance	F	% variance
Sequence length	2	22.6*	87.6	22.3*	85.8
Missing data	6	0.08	0.30	0.55	2.11
Rate of evolution	24	2.00†	7.76	1.94‡	7.47
Stochastic model	1	0.11	0.43	0.19	0.73

* Significant at $p < 0.001$.
† Significant at $p < 0.005$.
‡ Significant at $p < 0.01$.

Table 7.3. Examination of Additivity—Tukey Test

	F*	
Analysis	Unaltered Data	Transformed Data
Cladogram success	1.78×10^{-12}	8.07×10^{-12}
Cladogram resolution	5.19×10^{-6}	5.37×10^{-6}

*None of these values are significant at the $p = 0.05$ level.

number of taxa, sequence length, rate of evolution, and model of evolution. As for resolution, only sequence length and rate of evolution were significant. In neither case did missing data influence the phylogenetic outcome significantly.

The overwhelmingly important influence on the accuracy of the cladograms was the number of taxa, which accounted for 47% to 49% of the variance. The most important factor in the resolution of the cladograms was sequence length, responsible for 85 percent to 88 percent of the variance.

Four assumptions underlie any ANOVA procedure like this: normality, homoscedasticity, additivity, and independence. If the data do not conform to these prescriptions, the results of the analysis suffer accordingly. Although the method is somewhat robust to violations of the first two assumptions, both were checked through the use of the $\arcsin\sqrt{\theta}$ transformation of the data. This transformation tends to normalize the distribution of percents and fractions (normality) while equalizing differences in variance (homoscedasticity). Since the results were unchanged by the application of $\arcsin\sqrt{\theta}$, the likelihood of gross violation of either normality or homoscedasticity is low. The assumption of additivity—that the factors are linearly related to the outcome—was directly tested and upheld via the Tukey procedure. The final assumption of the method is independence (i.e., the factors are not inherently coupled in the determination of the cell values). Since the factors were varied externally by the experimental procedure, they are by their very nature independent. Thus the four assumptions appear to be met, and the variance in the data can be meaningfully partitioned among the factors mentioned earlier.

As shown in table 7.1, the number of taxa contributed the most to cladogram accuracy, accounting for almost one half of the variance. The second most important factor was the stochastic process or evolutionary model used to simulate the evolution of the sequences. Although the "constant" and "Poisson" model behaviors are similar, the Poisson model appears to reach a plateau, after which an increase in the number of characters fails to have the same beneficial effect. In the constant, or "clock," model of change, the effect from an increase in data is more linear, increasing with increasing numbers of characters. This linearity may explain the view of some investigators that doubling the length of sequences will double the accuracy of phylogenetic reconstruction—even with only a few taxa (Steele et al. 1991). If evolution were truly clocklike, such an argument would be more compelling.

Approximately one-sixth of the variation in cladogram accuracy can be ascribed to the length of the sequences, or the number of characters. As the sequences become longer (fig. 7.1), the phylogenetic hypotheses become more reliable (but the effect is not as pronounced in the 12-taxon case). A distant fourth among the factors influencing cladogram accuracy is the rate of evolution, the ratio of anagenetic to cladogenetic change. This ratio determines the extent of character change likely to occur between lineage-splitting events in the "evolution" of the modeled sequences. There is a slight decrease

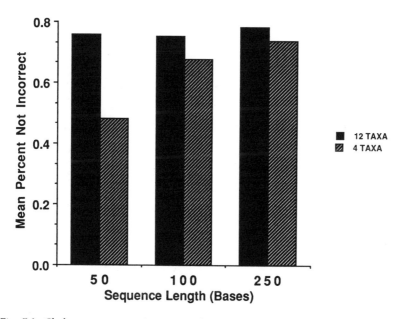

Fig. 7.1. Cladogram success (accuracy of reconstruction) versus sequence length. Although the increase in cladogram success for the complete data sets was not great, the culled data showed a large increment in success with increased numbers of characters.

in accuracy with very high and low rates of evolution. When rates are low, the sequences yield very little information about relationships among the taxa under study. When the rates are high, the sequences can approach randomness, containing little or no historical information. At both extremes, cladogram construction falters.

The second descriptor of cladogram behavior was resolution. The degree to which a cladogram was resolved offered a measure of its information content. If the scheme was completely resolved, whether correctly or incorrectly, it contributed a maximum of information on phylogenetic relationships. If it was completely unresolved, it yielded no information.

The factor of overwhelming importance for the resolution of the cladogram was sequence length. Almost 90% of the total variance was attributable to this one factor. Only the rate of evolution also contributed significantly, but to an order of magnitude less than the number of characters. The degree of resolution increased dramatically with longer sequences (fig. 7.2).

As with cladogram accuracy, the distorting effects of the rate of evolution

Cladogram Resolution vs. Sequence Length

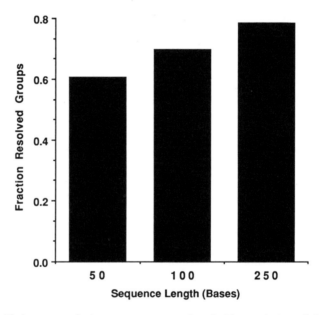

Fig. 7.2. Cladogram resolution versus sequence length. The resolution of cladograms increased markedly with increased number of characters.

Cladogram Resolution vs. Rate of Evolution

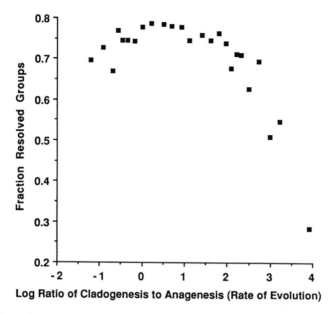

Fig. 7.3. Cladogram resolution versus rate of evolution. As with the accuracy of cladograms, the main effect of evolutionary rate comes in the extremely high and low values.

were most prominent at the very high and low ends of the range. When the rates were very low, cladograms were unresolved because of lack of sequence variation. At the highest rates, they were unresolved because of contradiction and confusion among the characters from the high number of hits (evolutionary events) at each position (fig. 7.3).

In neither of these analyses did missing data have a significant effect. In part, their lack of influence may stem from the way missing data were generated. There were two procedures by which missing data were introduced. The first assigned to each character an equal probability of becoming missing; thus the holes in the data matrix were completely random. In the second case, certain positions were more likely to go unobserved (missing) than others. This model reflected the circumstance in which certain characters are more difficult to observe, or are located in sequence areas of frequent insertion or deletion. But in no case were certain taxa (or specimens) more likely to contain missing data than others. The first two types of missing data approximate situations that arise often in molecular sequence data. The case of missing data varying by taxon, however, echoes situations involving fossil or extremely derived taxa and was not examined here. Nonetheless, if the results of this simulation are credible, missing data may not present as much of a problem in DNA studies as they do in morphologic ones.

Another problem not examined here is the effect of evolutionary rate on initial homology statements. If the sequences have undergone multiple hits at a high proportion of their positions, it can be very difficult to align these positions. Here the homologies were assumed to be known. In reality, homologies would not be known with absolute certainty, allowing corresponding adverse effects on the reconstruction of phylogenetic arrangements from sequence data.

These results are described with a final caution concerning their robustness. As stated earlier, the evolutionary model used was a significant factor in the analysis. Yet in modeling evolution, we choose from a world of possible distributions without knowing which (if any) are correct. How robust are these conclusions, relying as they do on specific models? It is hoped that effects of number of taxa, sequence length, and rates of evolution that seem to hold up under both models will endure.

These caveats aside, it seems clear that the accuracy of a cladogram is most influenced by the number of taxa, the area in which the effect of extinction exerts greatest influence. When only a few taxa are used, or when only a small number are available, the accuracy of phylogenetic reconstruction suffers. The inclusion of ever-longer sequences will alleviate the problem somewhat; but more likely, it will merely increase the level of detail (i.e., resolution) offered by incorrect statements. Accordingly, the inclusion of the highest number of taxa possible, in the long run, will be the best way to ensure accurate results. Presumably, all the taxa are not required to confidently reconstruct the essential framework of phylogenetic relationships. Nonetheless, studies that rely on the analysis of less than 1% of pertinent diversity must always be viewed in light of the effects of extinction, whether artificial or natural.

ACKNOWLEDGMENTS

I would like to thank Ranhy Bang, Elise Broach, Michael Novacek, Paul Vrana, and Quentin Wheeler for helpful comments on this manuscript. This research was supported by the Alfred P. Sloan Foundation.

REFERENCES

Donoghue, M. J., J. Doyle, J. Gauthier, A. Kluge, and T. Rowe. 1989. The importance of fossils in phylogeny reconstruction. *Ann. Rev. Ecol. Syst.* 20:431–460.

Farris, S. J. 1988. HENNIG 86, version 1.5.

Gardiner, B. 1982. Tetrapod classification. *Zool. J. Linnean Soc.* 74:207–232.

Gauthier, J., A. G. Kluge, and T. Rowe. 1988. Amniote phylogeny and the importance of fossils. *Cladistics* 4:105–209.

Patterson, C. 1981. The significance of fossils in determining evolutionary relationships. *Ann. Rev. Ecol. Syst.* 12:195–223.

Steele, K. P., K. E. Holsinger, R. K. Jansen, and D. W. Taylor. 1991. Assessing the reliability of 5S rRNA sequence data for phylogenetic analysis in green plants. *Mol. Biol. Evol.* 8:240–248.

Wheeler, W. C. 1989. The systematics of insect ribosomal DNA. In B. Fernholm, K. Bremer, and H. Jörnvall, eds., *The Hierarchy of Life*, pp. 307–321. Amsterdam: Elsevier.

8 : Measures of Phylogenetic Diversity

Kevin C. Nixon and Quentin D. Wheeler

Abstract. Phylogenetic diversity is defined as the relative species diversity of clades. Measurement of phylogenetic diversity requires minimal phylogenetic diagrams that summarize the relationships and species diversity of taxa under consideration. Two such measures are discussed. The unweighted measure consistently prioritizes all members of a clade as higher or lower than members of a particular related clade. In contrast, the weighted index allows some members of a clade to have different priorities relative to members of other clades, depending on the structure and degree of resolution within the clades under study. These measures are presented as a means of evaluating conservation priorities of sets of species, in order to identify those species that, if lost to extinction, would represent the greatest loss of phylogenetic diversity.

Biological diversity is declining at an unprecedented rate (Wilson 1985, 1988; Myers 1988; National Science Board 1989). It has been suggested that as many as half of all living species may become extinct within the next three decades (Lovejoy 1980; Raven 1988; Wheeler 1990). Given these projections, it is reasonable to conclude that humans will have the opportunity (responsibility) to make decisions that influence which species survive and which are relegated to extinction. By what criteria are such decisions to be made?

When possible, area conservation decisions should be made on the basis of careful considerations of overall species diversity, geography, vegetation, and habitat. Prioritization of such geographic areas, however, necessitates some kind of prioritization of their resident species. Some decisions will most certainly involve evaluation of single species on the basis of criteria of importance in the ecosystem, uniqueness, or human use. Do limited resources better go toward conservation of a carnivorous mammal, an owl, a fish, or an insect? The answer is deceptively complex, potentially involving some combination of factors related to the ecology, behavior, potential commercial value, genetic properties, and uniqueness of species.

Three major criteria for selection of areas and habitats for conservation have been formalized by the U.S. Congress, Office of Technology Assessment (1987): uniqueness of the ecosystem, genetic uniqueness of species and populations, and economic value of species. The second criterion, uniqueness of species, remains difficult to evaluate. Examples provided by OTA are "families with few species or genera with only one species." Such approaches to prioritizing species are not new and have been emphasized by various conservationists (Ehrlich 1988). Such determinations are sensitive to taxonomic classifications, where family and generic delimitation may vary widely in different groups of organisms. Instead of taxonomic criteria of uniqueness and diversity, conservation decisions should be based on criteria of *phylogenetic uniqueness* and *phylogenetic diversity*.

Phylogenetic diversity can be defined as the species richness of a clade. Phylogenetic diversity, although measured in absolute numbers of species, is more useful as a relative concept when species diversity of groups is compared. The placental mammals are phylogenetically diverse relative to the monotremes, assuming each is a monophyletic group. Without a phylogenetic perspective, groups are arbitrary (Farris 1983), and diversity comparisons have little, if any, meaning.

We define *phylogenetic uniqueness* as the phylogenetic (species) diversity of a monophyletic group relative to its sister group. The monotremes are phylogenetically unique, because they are phylogenetically equivalent to a large, diverse group of placentals and marsupials. In contrast, a small monophyletic group of three orchid species, the sister group of which has 10 species, is far less unique. Although these two examples are fairly simple, diversity and uniqueness are much more complex with the addition of more taxa and more phylogenetic resolution, as discussed later.

Most efforts at measuring biodiversity focus on species diversity, as viewed on a geographic or ecological basis. Conservation biologists often use the gap analysis concept to set priorities (Burley 1988). Gap analysis rests on the supposition that the preservation of major vegetative types will indirectly assure preservation of much of biological diversity. This approach does not support detailed, specific ranking of taxa, though assessments of vegetative and phytobotanical regions have been of paramount importance to date (Huntley 1988). Such measures of species diversity are extremely useful for determining broad priorities in the conservation of habitats that foster large numbers of species, such as tropical rain forests. However, measures of raw species diversity lack qualitative information about the phylogenetic relationships, and phylogenetic uniqueness, of the species involved. A statistic that embodies information about phylogenetic relationships and species diversity is appropriately termed a *phylogenetic diversity* index.

Recently, indices have been proposed that take into account aspects of phylogenetic relationship as the basis for setting priorities for the conservation of geographic areas (Vane-Wright et al. 1991; May 1990). Williams et al. (1991) have discussed a range of indices with similar aims. We do not propose to critique these measures in detail here, but simply observe that their application requires fully resolved cladograms and such cladograms exist for relatively few taxa. The two measures that we propose here do not require complete resolution of species relationships, and thus are more useful under most circumstances that conservationists are likely to encounter. It may be desirable to incorporate our phylogenetic diversity indices with biogeographic considerations, but this is not the intent of the present discussion. A more complete discussion of the indices presented here as contrasted with other measures is in preparation.

What Is Phylogenetic Diversity?

A hypothetical example will serve to illustrate the concept of phylogenetic diversity. Based solely on numbers of species, a group of five species of orchid has the same diversity as a group consisting of one species each of a mammal, fish, flowering plant, insect, and red alga. Intuitively, the five species of orchid represent less phylogenetic diversity than the eclectic group. In practice, most conservationists would likely agree to preserve the broader group of phylogenetic species before the five species of orchid. This would not be the case if the five orchid species were the last remaining extant orchids. To make a decision to preserve a broad spectrum of phylogenetic diversity we must also consider phylogenetic uniqueness. This example in no way diminishes the importance of preserving *all species,* but given limited resources, and the rapidity of the destruction of natural habitats, it may be necessary to assign priorities to the preservation of particular species. Certainly, it is impossible to place a scientific or monetary value on individual species. Each species is genetically unique and potentially phylogenetically informative. All or as many species and ecological systems as possible should be protected to the fullest extent. The current phenomenal rate of deforestation and of species extinction, however, obviates the ideal. When faced with preserving a limited number out of a set of endangered species, how can we set priorities to minimize diversity loss and maximize diversity preserved?

Current, formal listings of endangered and threatened species show that disproportionate weight is given to relatively few higher taxa, such as selected vertebrates. The composition of these lists does not necessarily reflect biological priorities but instead seems to be heavily influenced by an anthropocentric view of species importance. Such lists meet an important need but are inadequate and incomplete. A growing number of biologists believe that they need to be both taxonomically and phylogenetically representative of biodiversity.

Any measure of phylogenetic diversity or uniqueness should not rely on existing classifications, which vary greatly in application and often are at odds with phylogenetic information, but should be dependent on explicit hypotheses of phylogenetic relationship. We present here objective methods for estimating relative phylogenetic diversity of taxa (monophyletic groups of species). We propose two measures of phylogenetic diversity that can assist conservation biologists in setting priorities for the preservation of taxa.

The calculation of phylogenetic diversity should in no way diminish efforts to save habitats based on ecological and species diversity measures. Clearly, ecological considerations may provide insight into interactions among species that would identify "key" species that should be given priority for preservation because of potential detrimental effects on other species in the same ecosystems. However, a greater understanding of phylogenetic diversity will undoubtedly indicate phylogenetically unique species that should be given some priority for conservation efforts, as well as identify areas and habitats that harbor great phylogenetic diversity, which may not correspond directly to raw species diversity.

In terms of conservation, the goal of calculating phylogenetic diversity should be to maximize information about diversity of clades such that the phylogenetic uniqueness of particular species can be evaluated. Thus species belonging to large, diverse clades should have lower priority for conservation than species belonging to clades of low diversity. However, in essence, all extant organisms belong to *some* extremely large and diverse clades, and a measure of the rank order of clades is also necessary. Thus the monotypic plant genus *Ginkgo* probably represents the last surviving member of a clade that is at least the sister group of all living conifers, on the basis of shared derived features (Coulter and Chamberlain 1910; Doyle and Donoghue 1987). Based on its phylogenetic uniqueness, *Ginkgo* should have a higher priority for conservation than any *particular species* of pine or fir, since the "conifer clade" is relatively diverse. Although several extinct forms of *Ginkgo* are known from the Tertiary of North America, the past diversity of the clade is not relevant to its status as a clade equal to all conifers or to the lack of extant diversity in the *Ginkgo* clade. Taxa of similar importance are identified easily. For example, Zygentoma, with about 320 species, is the sister group of remaining insects with potentially millions of living species, and Archaeognatha, with 250 species, is the sister group of these combined (Hennig 1981; Kristensen 1975, 1981; Wilson 1985; Wheeler 1990). Other examples are the monotremes relative to the marsupials or the marsupials in relation to placental mammals. It should be noted here that any measure of phylogenetic diversity must be relative. In other words, the diversity of a group must be evaluated relative to sister groups, not in absolute terms.

The preceding examples, such as *Ginkgo,* reflect the desirability of preserving phylogenetic uniqueness. Another aspect of conservation efforts should be to maximize the overall diversity that is preserved. If 50 species of *Ginkgo* were still extant, should all 50 be given conservation priority over any particular conifer species? In a certain sense, the preservation of uniqueness and overall diversity are at odds. Our intent is not to answer the question of whether to favor conservation of overall diversity or conservation of uniqueness. The interplay between these factors will become clearer in our discussion of two different methods of ranking species in terms of phylogenetic diversity.

Two Measures of Phylogenetic Diversity

We have developed two preliminary measures of phylogenetic diversity that can be used to rank species according to the amount of phylogenetic diversity in the clades to which they belong. The rankings produced by these methods, if used as conservation priorities, are consistent with the preservation of rare phylogenetic entities. In both measures, a lower value indicates that the species is more phylogenetically unique, so the ranks indicated range from highest priority of 1 to a rank equal to the total number of species evaluated. We shall call these the (unweighted) binary phylogenetic diversity index and the weighted phylogenetic diversity index.

The unweighted index is calculated by comparing species diversity of sister clades, beginning at the base of a cladogram, and for each level giving a

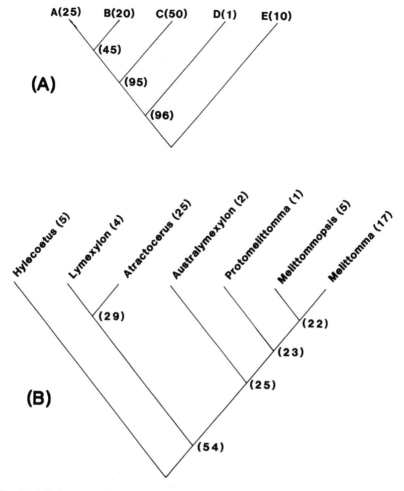

Fig. 8.1. Phylogenetic diversity indices: (A) Hypothetical example with taxa A–E; (B) example from genera of Lymexylidae (Coleoptera). Numbers of species shown in parentheses for each terminal group and for the component groups represented by the internal nodes.

binary score of 1 to the most diverse clade and 0 to the least diverse clade (its sister clade). At unresolved nodes, additional levels are introduced as necessary to distinguish among clades, so that less diverse clades are given lower scores.

The unweighted phylogenetic diversity index is illustrated in a simple example in Figure 8.1A. Each term of the cladogram has a value equivalent

Table 8.1. (Unweighted) Binary Rank Phylogenetic Diversity Index

Node	1	2	3	4		Decimal	Rank Order
BPD(A) =	1	1	0	1	=	13	4
BPD(B) =	1	1	0	0	=	12	3
BPD(C) =	1	1	1	—	=	14	5
BPD(D) =	1	0	—	—	=	8	2
BPD(E) =	0	—	—	—	=	0	1

to the number of included species, and each node a value summing the number of species included in all clades descended from the node. In this example the unweighted binary phylogenetic diversity (BPD) of each terminal component is calculated as in table 8.1.

If a node does not subtend a species, that species is scored with a dash in the preceding matrix, but in calculating the rank any dashes are read as 0. Thus the score for each species may be read as a bitwise representation of a binary number (e.g., $1101 = 13$ decimal). If one prefers, the binary (base 2) numbers may be read directly as decimal integers (e.g., 1011, 1100, 1110, 1000, and 0) with the same resulting rank orders. Viewing the scores directly as decimal representations is easier in large cladograms with many nodes, where binary numbers could get huge.

The unweighted binary index will rank all members of a clade as having the same priority *relative* to all members of its sister clade. Thus in a classification based on phylogeny, when comparing two genera, all species of one genus will have a higher rank than all species of the other genus, although the species within each genus may have different ranks relative to each other. Although this measure is probably the best estimate of relative phylogenetic uniqueness, it does not disperse the rankings so that the *overall* sample of phylogenetic diversity is maximized. For example, using the unweighted measure, all orchid species might be ranked lower than all species of liverwort. There can exist no measure that maximizes both uniqueness and breadth of sample, but the following weighted index is perhaps the best compromise of these two goals. The weighted index is calculated as follows:

$$\text{WPD} = \sum_{1}^{n} d(i)$$

where $d(i)$ is diversity measured in number of species for each group i to which the species belongs. The index produced by such an analysis will vary from a minimum value of 0 (the species is the last remaining sister species of all other species considered) to very high values when species from widely separated speciose groups are being considered.

Table 8.2. Weighted Phylogenetic Diversity Index

							Rank Order
WPD(A) =	25 +	45 +	95 +	96	=	261	5
WPD(B) =	20 +	45 +	95 +	96	=	256	4
WPD(C) =		50 +	95 +	96	=	241	3
WPD(D) =			1 +	96	=	97	2
WPD(E) =				10	=	10	1

Using the example in figure 8.1A, the weighted phylogenetic diversity of each species and the rank order of each species are calculated as in table 8.2. Note that the ranking using the weighted index is different from that obtained with the unweighted index. In this case, the two species from groups A and B, similar in size to that of group C, are given lower rankings. Thus instead of prioritizing two species of one group over a single species of its sister group, the rankings create a more balanced sample of the diversity present in the cladogram.

The weighted index presented earlier can also be viewed in a nonphylogenetic context, as a weighted index of hierarchic diversity. Thus it may be useful for determining sampling strategy for any hierarchically ordered set of objects, where numbers of objects in each nested set are to be considered. Depending on the goals of such sampling, hierarchic diversity could be maximized or minimized (the latter being analogous to favoring phylogenetic uniqueness).

The weighted phylogenetic diversity index presented here is relative, in that it will only rank a particular set of species under consideration and will not produce an absolute measure of phylogenetic diversity. Thus if other species come under consideration, the topology of the cladogram will become more detailed, and rankings of species may change. This is not necessarily inconsistent with the goals of conservation priorities, as priorities change when particular species and habitats become threatened.

The weighted index of phylogenetic diversity provides a way of estimating the status of higher-level groups, or clades. Priorities based on such estimates are intuitively appealing, since the loss of the last member of a clade should be considered a greater phylogenetic loss than the loss of a single species of its more diverse sister clade.

Other components of phylogenetic pattern, such as character distance (patristic distance on a cladogram), might also be considered as the basis for an index of phylogenetic diversity (see Williams et al. 1991). Indices based on such measures are sensitive to the data source, method of sampling, and completeness of our knowledge to a much greater extent than our index based on phylogenetic topology. Although the topologically based indices we present here are sensitive to changes in our understanding of phylogenetic relationships and species, they are not sensitive to data sources or sampling per se.

Analysis of a Complete Phylogeny: Lymexylidae

Using either of the two phylogenetic diversity indices, it is possible to rank the genera of ship-timber beetles (Coleoptera: Lymexylidae) according to their relative value as components of phylogenetic diversity in the family. The family includes seven genera classified in three subfamilies by Wheeler (1986): Hylecoetinae (*Hylecoetus* Linnaeus); Lymexylinae (*Lymexylon* Fabricius, *Atractocerus* Palisot de Beauvois); and Melittomminae (*Australymexylon* Wheeler, *Protomelittomma* Wheeler, *Melittommopsis* Lane, and *Melittomma* Murray). The cladogram in figure 8.1B shows the relative relationships among the genera of Lymexylidae and summarizes the number of species within each genus. The additive number of species within each clade is indicated by the parenthetical number at each node of the cladogram. The two phylogenetic diversity indices, then, can be calculated as in table 8.3.

Thus, based on the weighted index, the genera of lymexylids might be

Table 8.3. Unweighted Binary Phylogenetic Diversity

Node	1	2	3	4	5	6	Rank
BPD (*Hylecoetus*) =	0	0	—	—	—	—	1
BPD (*Lymexylon*) =	1	1	0	—	—	—	6
BPD (*Atractocerus*) =	1	1	1	—	—	—	7
BPD (*Australymexylon*) =	1	0	0	0	—	—	2
BPD (*Protomelittomma*) =	1	0	0	1	0	—	3
BPD (*Melittommopsis*) =	1	0	0	1	1	0	4
BPD (*Melittomma*) =	1	0	0	1	1	1	5

Weighted Phylogenetic Diversity

							Rank
WPD (*Hylecoetus*) =					5 =	5	1
WPD (*Lymexylon*) =				4 +	29 +	54 =	87	3
WPD (*Atractocerus*) =				25 +	29 +	54 =	108	5
WPD (*Australymexylon*) =				2 +	25 +	54 =	81	2
WPD (*Protomelittomma*) =			1 +	23 +	25 +	54 =	103	4
WPD (*Melittommopsis*) =	5 +	22 +	23 +	25 +	54 =	129	6
WPD (*Melittomma*) =	17 +	22 +	23 +	25 +	54 =	141	7

prioritized as follows: *Hylecoetus, Australymexylon, Lymexylon, Protomelittomma, Atractocerus, Melittommopsis, Melittomma*. This is in marked contrast to the unweighted index, which prioritized the species on a clade-by-clade basis. In general, the weighted index produces a broader sampling of the phylogenetic diversity present in a cladogram, since the adjacent ranks tend to alternate back and forth among sister clades, a phenomenon that cannot occur using the unweighted binary index.

Analysis of Selected Species: Fagaceae

The indices of relative phylogenetic diversity may be applied to incomplete sets of taxa, so long as the relative phylogenetic relationships among them are known or hypothesized and the number of excluded taxa (at terms or along internodes) are known or can be reasonably estimated. Such applications for the phylogenetic diversity indices can be illustrated by an examination of several selected taxa from Fagaceae, each of which is or may conceivably become the object of potential conservation efforts. In this hypothetical example, five fagaceous species are under consideration: *Colombobalanus excelsa* (Lozano et al. Nixon and Crepet, *Formanodendron doichangensis* (Camus) Nixon and Crepet, *Quercus acerifolia* Stoynoff and Hess, *Q. tomentella* Greene, and *Q. brandegei* Goldman. The relative phylogenetic relationships, based in part on Nixon and Crepet (1989), are presented in figure 8.2. All the species indicated are members of the "quercoid" clade of subfamily Fagoideae. Numerous other species of *Quercus* occur that must be accounted for in this analysis, even though the cladogram does not resolve all species. Thus very rough numbers of species (rounded within 50 species for simplicity of calculation) for each clade are indicated at each node on the diagram, accounting for all species that would be included above that node in a more detailed cladogram. Even though the cladogram is resolved only to show the relative relationships of the five species under consideration, the diversity of clades not shown is still used in the calculation (see table 8.4). In this case, the weighted and unweighted diversity indices result in the same relative rankings. Based on the taxa that have the lowest modern diversity, the priority for conservation would be in the same order as that presented in table 8.4.

Some possible effects of adding additional species to an incomplete analysis are illustrated in figure 8.3. In general, as the number of species considered in a clade increases, the weighted index for any particular species also increases. The amount of increase is related to the particular resolution of species relationships, in combination with the number of species represented by each subclade. Four different hypothetical examples are presented. In

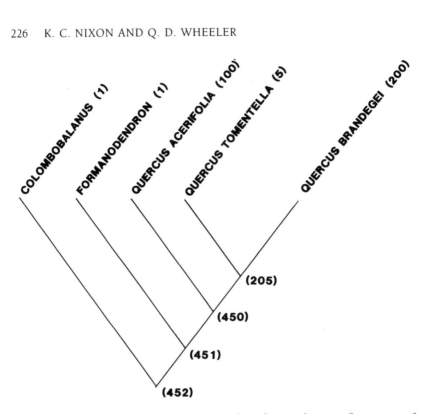

Fig. 8.2. Phylogenetic diversity indices. Hypothetical example using five species of Fagaceae. Numbers in parentheses represent very rough estimates of numbers of species represented at each node and terminal branch. Nodes include groups not included as terminals, and thus have higher values.

three of these cases, we consider 10 (hypothetical) species of red oak instead of the single species *Q. acerifolia*. In the last case, we consider 11 species. In each of these cases, the BPD produces the same relative ranks as the WPD, and the values for the BPD are not presented.

> *Equally diverse, unresolved subclades:* In the first example (see fig. 8.3A), we possess no special knowledge of the relationships of the 10 red oak species, so the red oak clade is completely unresolved. We have arbitrarily created a balanced example, in which the 10 species are distributed among 10 equally sized subclades of 10 species each. The WPD for each of the 10 red oaks in this example is 1011, and we are not able to discriminate among them. If group B were a completely unresolved bush of 100 separate species, without internal resolution, each species would have the same score as *Q. acerifolia* in the original example.

Table 8.4. Unweighted Binary Phylogenetic Diversity Index

	1	2	3	4	Rank
BPD (*Colombobalanus*)	0	—	—	—	1
BPD (*Formanodendron*)	1	0	—	—	2
BPD (*Quercus acerifolia*)	1	1	0	—	3
BPD (*Q. tomentella*)	1	1	1	0	4
BPD (*Q. brandegei*)	1	1	1	1	5

Weighted Phylogenetic Diversity Index

					Score	Rank
WPD (*Colombobalanus*)					0	0
WPD (*Formanodendron*)				0+	451	451
WPD (*Quercus acerifolia*)			100+	450+	451	1001
WPD (*Q. tomentella*)		5+	205+	450+	451	1111
WPD (*Q. brandegei*)	250+	205+	205+	450+	451	1356

Equally diverse, completely resolved, pectinate: In figure 8.3B, the sub-clades remain equally diverse with 10 species each, but the relation-ships among these subclades are fully resolved and pectinate. In such cases, the lowest WPDs are for species belonging to "basal" subclades, with the highest scores for species in clades that are nested higher in the pectinate cladogram.

Equally diverse, balanced, partially resolved: In this example (see fig. 8.3C), the species are distributed among 10 subclades of 10 species each, and we have phylogenetic resolution that indicates two equal groups of five subclades. Again, since the distribution of species diversity and cladistic resolution is balanced, all 10 species in the example have the same WPD (1061).

Unequally diverse, partially resolved: In this example (see fig. 8.3D), the distribution of species is unequal, with 11 subclades. The single species designated A is the last remaining member of the sister clade to the remainder of the species. The score for the single species (A) is 1001, identical to the score if it had been the only species con-sidered in the whole group of 100. The other species in the unre-solved subclade have higher scores, 1109 for species C and 1110 for all other species.

These examples illustrate several points about the WPD. One is that when a single species from a clade is considered, it will receive a score that is equivalent to the lowest score possible for a species if the clade were fully

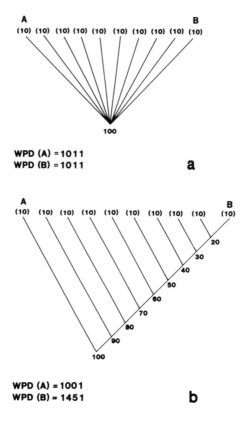

a

b

resolved internally. Such a score would occur within a fully resolved clade *only* if a single species were the last member of a sister group of all remaining species (cf. species B1 in fig. 8.3D and *Q. acerifolia* in fig. 8.2). The WPD therefore is a conservative estimate of diversity—and in unresolved situations will assume that any clade is a "basal" or sister-group clade of the remaining species. We feel that this bias of the WPD is consistent with conservation goals, since the benefit of the doubt is given when phylogenetic resolution is not available.

Another aspect of the WPD illustrated in these examples is that lack of resolution among equally diverse clades results in tied scores. Such ties should be distinguished from ties that are coincidental and not due to unresolved clades. In the case of coincidental ties in WPD scores, other criteria must be used to assess priority. In the case of ties due to lack of resolution, anomalies in the distribution of WPDs may occur. An example showing the possible problems inherent in unresolved ties is presented in figure 8.4.

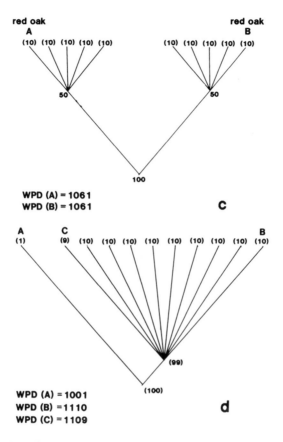

Fig. 8.3. Hypothetical extension of example in figure 8.2. WPDs are calculated assuming the subtrees presented in a–d are attached in place of the "red oak" clade (*Q. acerifolia* branch in fig. 8.2). (a) Red oak clade with hypothetical subgroups of 10 species each, completely unresolved; (b) red oak group with 10 subclades, resolved as a pectinate tree; (c) red oak clade with two equal groups of five unresolved subgroups; (d) red oak clade with partial resolution and unequal subgroups.

Consider three hypothetical clades, as depicted in figure 8.4. In the first example, any single species drawn from clade B will have WPD = 60. The same score would be determined if a completely unresolved phylogeny of the five species of clade B were hypothesized. Also, the same score would be determined for a species that was the last remaining species of the sister clade to the other four species, as species B1 in fig. 8.4B. In fig. 8.4B, we also see that the highest possible score that could occur for a species drawn from the clade of five species would be 69, as calculated for B4 and B5. In this

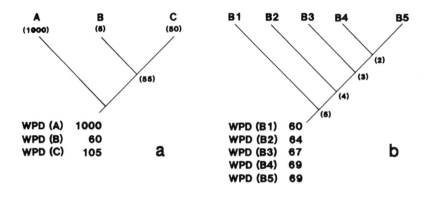

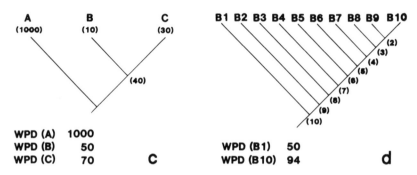

Fig. 8.4. Hypothetical example showing effects of unresolved groups on rankings: (a) cladogram in which three species are considered; (b) pectinate resolution within group B from cladogram of a; (c) cladogram in which three species are considered; (d) pectinate resolution within group B from cladogram C.

example, all five species of clade B would receive lower WPDs, and therefore have higher priority for conservation, than any species from clade A or C. Thus the lack of resolution implied in the B clade in figure 8.4A has not created any problems other than the inability to distinguish among a cluster of species.

In contrast to the previous example, figure 8.4C illustrates a situation in which lack of resolution may actually mask potential ranking differences in WPD scores. This example is parallel to that of figure 8.4A and B, but the possible scores among the 10 species of clade B have a range that overlaps the value for the species drawn from clade C. In such a case, if the 10 species were considered equal, with a score of 50, as they would be if no resolution within clade B were recognized (fig. 8.4C), one might be tempted to prioritize all 10 species of B over the single species of clade C. This would be incorrect. The range of scores for species of clade B, when clade B is fully resolved, can range from 50 to 94. Thus some of the species of clade B would in fact receive a higher WPD score (and lower priority) than the single species of clade C. In practice, this would imply that a threshold WPD score exists, equivalent in this example to the score for any single species of clade C, such that a WPD score in excess of this value would suggest that species of clade B should receive a relatively lower ranking in conservation decisions.

Phylogenetic Diversity and Conservation Biology

All species are not equally "unique" in a phylogenetic sense. Given any significant degree of phylogenetic resolution for a clade, some constituent species will be assigned logically a higher priority for conservation than others. This is due to the phylogenetic relationships among subclades and the distribution of species among these subclades.

How species are prioritized depends to a great extent on the stated goals for conservation. If estimates of species extinction rates are approximately correct, then the fundamental decisions do not relate to whether we conserve species or not but instead to which half of the species might survive into the future. Given the diverse reasons for wanting to preserve biological diversity, it seems reasonable to conclude that responsible conservation will take into account the phylogenetic relationships among those species maintained in natural and/or artificial habitats. Educated, specific decisions will therefore depend upon the application of one or more measures of such phylogenetic diversity.

The measures presented here permit the relative ranking of taxa based on the results of basic taxonomic and cladistic research. We do not suggest that these or other conceivable measures of phylogenetic diversity be used in isolation to make conservation decisions. Phylogenetic diversity is instead an additional source of information to guide such policy-making. Because of the existing paucity of both species-inventory and phylogenetic data, measures such as ours can be applied only in isolated ways. It is impossible to assess entire floras or faunas, or to compare the totality of phylogenetic distinctiveness between geographic areas or forest types.

Vane-Wright et al. (1991) discussed how, utilizing phylogenetic data, areas might be prioritized, including an example based on bumble bee data. Williams et al. 1991 discuss half a dozen similar measures, offering a range of options for making such weighting schemes. Although we must ultimately make decisions about which and how many natural habitats to maintain in perpetuity, existing taxonomic and phylogenetic data make it unlikely that these kinds of decisions can be made responsibly unless more precise goals and criteria are developed. A good decision relative to one taxon may be—and in all likelihood will be—a less optimal decision for another taxon. Maintenance of all taxa would likely encompass the preservation of the entire globe, an impossible task. Unfortunately, the unattainability of this optimal solution suggests the need for phylogenetic measures in the first place.

We have presented two simple measures of phylogenetic diversity that can be used as a guide for the establishment of conservation priorities. To apply these indices, one must have some estimate of the phylogenetic relationships of the species considered and the number of species in each clade of the minimally resolved cladogram. These indices are relatively easy to apply to circumstances where decisions must be made to collect, study, or conserve one or another taxon.

The unweighted binary index has the advantage of being stable in relative rankings even when additional species are considered. Thus unless our understanding of the number of species or actual phylogenetic relationships changes, all orchids will always have the same priority relative to all whales. The unweighted index is unrelenting in favoring phylogenetic uniqueness in the assignment of priority. Given sister clades of very similar species diversity, the binary index will give higher priority to all species in the smaller clade than to any species in the larger clade. Thus the unweighted index is the best measure of relative phylogenetic diversity and uniqueness. As such it is more predictable and stable than the weighted index, but may not be ideal for making conservation decisions.

The weighted index is a compromise between phylogenetic uniqueness and sampling a broader range of phylogenetic diversity. It is unlikely, with the weighted index, that all members of one clade will have priority over all members of a sister clade of similar but slightly higher species diversity. This propensity of the weighted index to alternate priorities among species in sister clades, as resolution in each clade increases, provides a more balanced ranking for priority decisions. For this reason, we favor the weighted index when making priority decisions about individual species. The unweighted index may be more appropriate when considering geographic or area priorities. It might be used by ranking all species at the sites being considered, then scoring each site for a composite or average rank. This would provide an estimates of the relative phylogenetic uniqueness of the sites.

We hope that these and similar measures will continue to be developed to assist in making intelligent and responsible conservation decisions. We stress, however, that any phylogenetic measure depends upon the availability of sound and comprehensive data, minimally including complete inventories of existing species, comprehensive taxonomic treatments of the taxa, and availability of credible, corroborated cladistic hypotheses of relationships.

ACKNOWLEDGMENTS

We thank James Carpenter, William Crepet, Jerrold Davis, Melissa Luckow, and Michael Novacek for discussion of various aspects of the topics presented here. We especially thank Dr. C. J. Humphries for providing an advanced copy of the Williams et al. (1991) manuscript for our consideration.

REFERENCES

Burley, F. W. 1988. Monitoring biological diversity for setting priorities in conservation. In E. O. Wilson, ed., *Biodiversity*, 227–34. Washington, D.C.: National Academy of Science Press.

Coulter, J. M. and C. J. Chamberlain. 1910. *Morphology of the Gymnosperms*. Chicago: University of Chicago Press.

Doyle, J. A. and M. J. Donoghue. 1987. The origin of angiosperms: A cladistic approach. In E. M. Friis, W. G. Chaloner, and P. R. Crane, eds., *The Origins of Angiosperms and Their Biological Consequences*, pp. 17–50. Cambridge: Cambridge University Press.

Ehrlich, P. R. 1988. The loss of diversity: causes and consequences. In E. O. Wilson, ed., *Biodiversity*, pp. 21–27. Washington, D.C.: National Academy of Science Press.

Farris, J. S. 1983. The logical basis of phylogenetic analysis. In N. I. Platnick and V. A. Funk, eds., *Advances in Cladistics*, vol. 2, pp. 7–36. New York: Columbia University Press.

Hennig, W. 1981. *Insect Phylogeny*. London: Academic Press.

Huntley, B. J. 1988. Conserving and monitoring biotic diversity: Some African examples. In E. O. Wilson, ed., *Biodiversity*, pp. 248–260. Washington, D.C.: National Academy of Science Press.

Kristensen, N. P. 1975. The phylogeny of hexapod "orders". A critical review of recent accounts. *Z. Zool. Syst. Evol. Forsch.* 13:1–44.

Kristensen, N. P. 1981. Phylogeny of insect orders. *Ann. Rev. Entomol.* 26:135–157.

Lovejoy, T. E. 1980. A projection of species extinctions. In G. O. Barney, ed., *The Global 2000 Report to the President Entering the 21st Century*, vol. 2. Washington, D.C.: Council on Environmental Quality, U.S. Government Printing Office.

May, R. M. 1990. Taxonomy as destiny. *Nature* 347:129–130.

Myers, N. 1988. Tropical forests and their species: Going, going . . . ? In E. O. Wilson, ed., *Biodiversity*, pp. 28–35. Washington, D.C.: National Academy of Science Press.

National Science Board. 1989. *Loss of Biological Diversity: A Global Crisis Requiring International Solution*. Washington, D.C.: U.S. National Science Foundation.

Nixon, K. C. and W. L. Crepet. 1989. *Trigonobalanus* (Fagaceae): Taxonomic status and phylogenetic relationships. *Amer. J. Bot.* 76:828–841.

Raven, P. H. 1988. Our diminishing tropical forests. In E. O. Wilson, ed., *Biodiversity*, pp. 119–122. Washington, D.C.: National Academy of Science Press.

U.S. Congress, Office of Technology Assessment. 1987. *Technologies to Maintain Biological Diversity.* Washington, D.C.: U.S. Government Printing Office.

Vane-Wright, R. I., C. J. Humphries, and P. H. Williams. 1991. What to protect?—Systematics and the agony of choice. *Biol. Conserv.* 55:235–254.

Wheeler, Q. D. 1986. Revision of the genera of Lymexylidae (Coleoptera: Cucujiformia). *Bull. Amer. Mus. Nat. Hist.* 183:113–210.

Wheeler, Q. D. 1990. Insect diversity and cladistic constraints. *Ann. Entomol. Soc. Amer.* 83:1031–1047.

Williams, P. H., C. J. Humphries, and R. I. Vane-Wright. 1991. Salvaging diversity: divergent taxonomic measures for conservationists. In press.

Wilson, E. O. 1985. The biological diversity crisis: A challenge to science. *Issues Sci. Technol. Fall:* 20–29.

Wilson, E. O., ed. 1988. *Biodiversity.* Washington, D.C.: National Academy of Science Press.

List of Contributors

Gregory D. Edgecombe
Department of Geology
University of Alberta
Edmonton, Alberta, Canada T6G 2E3

David A. Grimaldi
Department of Entomology
American Museum of Natural History
New York, NY 10024

Bruce MacFadden
Florida Museum of Natural History
University of Florida
Department of Natural Sciences
Gainesville, FL 32611

Kevin C. Nixon
Bailey Hortorium
Cornell University
Ithaca, NY 14853

Mark A. Norell
Department of Vertebrate Paleontology
American Museum of Natural History
New York, NY 10024

Michael J. Novacek
Department of Vertebrate Paleontology
American Museum of Natural History
New York, NY 10024

Quentin D. Wheeler
Department of Entomology
Cornell University
Ithaca, NY 14853

Ward C. Wheeler
Department of Invertebrates
American Museum of Natural History
New York, NY 10024

Index

Italics indicates illustrations, diagrams, or tables.